# 闻雪友文集

闻雪友 著

# 内 容 简 介

本书是闻雪友院士多年从事燃气轮机科研工作的理论研究成果,涵盖中英文科技文章30余篇。文集内容方向为介绍燃气轮机方面的理论成果,具体包括燃气轮机高温叶中表面温度测量、蒸汽回注燃气轮机技术以及船用燃机、船用柴油机等内容。

本文集反映了闻雪友院士在燃气轮机领域做出的学术贡献,可供燃气轮机行业的科研人员、技术人员参考,也可供高等学校燃气轮机专业师生参阅。

**图书在版编目(CIP)数据**

闻雪友文集 / 闻雪友著. —哈尔滨:哈尔滨工程大学出版社,2020.1
ISBN 978-7-5661-2186-8

Ⅰ.①闻… Ⅱ.①闻… Ⅲ.①燃气轮机-文集 Ⅳ.①TK47-53

中国版本图书馆 CIP 数据核字(2019)第 290710 号

**选题策划:**卢尚坤 薛 力
**责任编辑:**卢尚坤 张 昕
**封面设计:**李海波

出版发行 哈尔滨工程大学出版社
社　　址 哈尔滨市南岗区南通大街 145 号
邮政编码 150001
发行电话 0451-82519328
传　　真 0451-82519699
经　　销 新华书店
印　　刷 北京中石油彩色印刷有限责任公司
开　　本 787 mm×1 092 mm　1/16
印　　张 19.25
字　　数 431 千字
版　　次 2020 年 1 月第 1 版
印　　次 2020 年 1 月第 1 次印刷
定　　价 98.00 元

http://www.hrbeupress.com
E-mail:heupress@hrbeu.edu.cn

# 《中国工程院院士文集》总序

二〇一二年暮秋,中国工程院开始组织并陆续出版《中国工程院院士文集》(简称《文集》)系列丛书。《中国工程院院士文集》收录了院士的传略、学术论著、中外论文及其目录、讲话文稿与科普作品等。其中既有早年初涉工程科技领域的学术论文,亦有成为学科领军人物后,学术观点日趋成熟的思想硕果。卷卷《文集》在手,众多院士数十载辛勤耕耘的学术人生跃然纸上,透过严谨的工程科技论文,院士笑谈宏论的生动形象历历在目。

中国工程院是中国工程科学技术界的最高荣誉性、咨询性学术机构,由院士组成,致力于促进工程科学技术事业的发展。作为工程科学技术方面的领军人物,院士们在各自的研究领域具有极高的学术造诣,为我国工程科技事业发展做出了重大的、创造性的成就和贡献。《中国工程院院士文集》既是院士们一生事业成果的凝练,也是他们高尚人格情操的写照。能为中国工程院出版史上留下这样丰富深刻的一笔,余有荣焉。

我向来以为,为中国工程院院士们组织出版《中国工程院院士文集》之意义,贵在"真善美"三字。他们脚踏实地,放眼未来,自朴实的工程技术升华至引领学术前沿的至高境界,此谓其"真";他们热爱祖国,提携后进,具有坚定的理想信念和高尚的人格魅力,此谓其"善";他们治学严谨,著作等身,求真务实,科学创新,此谓其"美"。《中国工程院院士文集》集真善美于一体,辩而不华,质而不俚,既有"居高声自远"之澹泊意蕴,又有"大济于苍生"之战略胸怀,斯人斯事,斯情斯志,令人阅后难忘。

读一本《文集》,犹如阅读一段院士的攀登高峰的人生。让我们翻开《中国工程院院士文集》,进入院士们的学术世界。愿后之览者,亦有感于斯文,体味院士们的学术历程。

徐匡迪

二〇一二年

# 出 版 前 言

燃气轮机作为装备制造业皇冠上的明珠，承载了我国能源与动力领域一代又一代人的梦想与努力。七十年来，在党中央和各级政府的关怀下，燃气轮机的开拓者们筚路蓝缕、艰苦拼搏，一代又一代的继任者锐意进取、创新发展，燃气轮机行业在经历了探索尝试、引进仿制、自主研发几个重要阶段的磨砺后，依托国家"航空发动机和燃气轮机"重大科技专项，初步建立了自主创新的基础研究、技术与产品研发和产业体系。

闻雪友院士长期从事舰船及工业燃气轮机的研究设计工作。其曾任我国第一台航空改装大功率舰船燃气轮机的技术负责人、我国第一台第二代舰船燃气轮机的技术负责人、新型国产化舰船燃气轮机的总设计师，为我国舰船动力现代化做出突出贡献。在能源动力工程方面，闻雪友院士在国内首先研究建成双工质平行复合循环电站，并推广应用；担任"国家高技术研究发展计划（简称"863 计划"）"10 MW 高温气冷反应实验堆二期工程中我国首次研制的氦气透平压气机组子项目的总设计师；获全国科学大会奖、国家科学技术进步奖、国防科学技术奖、军队科技进步奖等多项殊荣。

闻雪友院士长期从事能源与动力领域的研究工作，在燃气轮机领域做出了突出的贡献，在科研生涯中将自己的科研成果结晶为多篇学术论文并在国内外期刊上发表。此次将院士多年杰出的理论研究成果整理汇编成书。

本书出版历时较长，对内容和结构进行了多次调整和补充。但遗憾的是，部分文章因为时间久远，相关数据、内容及文献已经遗失或不可考，只能留给读者自己去思考。

此次出版的《闻雪友文集》，大略按内容编次，仅进行勘误、注释等简单的编辑工作，最大限度地保留了每篇文章的原貌。希望借此书能将优秀的理论成果更广泛地传承和发扬，以期为相关领域的发展再次助力。

哈尔滨工程大学出版社
2020 年 1 月

# 目 录

## 一、燃气轮机高温叶中表面温度测量

依靠技术优势 开拓国内外市场 ········································· 3
双工质平行 - 复合循环发动机在船上应用的可能性 ······················· 7
日本新一代船用燃气轮机(SMGT)研究计划 ······························· 16
国产航空斯贝发动机的工业及船用化发展 ······························· 23

## 二、蒸汽回注燃气轮机技术

M70 系列燃气轮机的发展与应用 ······································· 35
对发展大功率船用燃气轮机的新思考 ··································· 42
现代舰船燃气轮机发展趋势分析 ······································· 51
船用燃气轮机 GT1000 ··············································· 59
WR - 21——新一代的船用燃气轮机 ···································· 65
舰船燃气和蒸汽动力装置的发展与展望 ································· 75
双工质平行 - 复合循环热机(程氏循环热机) I ··························· 81
双工质平行 - 复合循环热机(程氏循环热机) II ·························· 88
燃气轮机 STIG 化的研究 ············································· 99
燃气轮机回注蒸汽装置的研究 ········································· 107
燃气轮机湿空气回注循环分析 ········································· 113
部分回热回注蒸汽燃气轮机循环的研究 ································· 121
应用于 PRSTIG 循环化 SIA - 02 燃气轮机组上的喷射器 ·················· 127
柴油机注汽涡轮增压系统 ············································· 131
PG5361 STIG 装置 ·················································· 137

M1A-01-CC 程氏循环热电联供装置 …………………………………………………… 142

STIG 的新进展 …………………………………………………………………………… 147

燃气轮机余热锅炉市场趋向 …………………………………………………………… 151

涡轮转子整圈带冠叶片振动分析 ……………………………………………………… 155

涡轮导向器面积调整对燃气轮机性能的影响 ………………………………………… 162

船用燃气轮机箱装体板壁隔声设计 …………………………………………………… 177

带膜盘联轴器的轴系临界转速分析 …………………………………………………… 188

## 三、ASME 上发表的文章

Feasibility Study of an Intercooled-cycle Marine Gas Turbine …………………… 203

Experimental Study on Influence of Cooling Air Flow upon Surface Temperature of High-temperature Turbine Blades in Gas turbine ……………………………… 214

An Adjustment Method of Axial Force on Marine Multi-shaft Gas Turbine Rotor …… 223

An Evaluation of the Application of Nanofluids in Intercooled Cycle Marine Gas Turbine Intercooler ……………………………………………………………………… 234

An Alignment Monitoring Device for the Output Shafting of Marine Gas Turbine …… 257

Industrial and Marine Development Policy Study and Practices for GT28 Gas Turbine …… 264

## 四、其他(船用燃机、船用柴油机)

EGT 在燃气轮机燃用灰分燃料方面的经验 …………………………………………… 285

苏联船用燃气轮机的摇篮——МАШПРОЕКТ ………………………………………… 297

# 一、燃气轮机高温叶中表面温度测量

# 依靠技术优势　开拓国内外市场*

闻雪友

（中国船舶重工集团公司第七〇三研究所）

　　七〇三所是中国船舶工业总公司第七研究院下属的一个舰船动力设备研究所,组建于1961年6月。其拥有燃气轮机、蒸汽轮机及热能工程、船舶锅炉及工业锅炉、自动控制等11个研究室,1个试制工厂和1个以高新技术为主的科工贸总公司。

　　研究所已研制舰艇用多型蒸汽轮机主动力装置、燃气轮机主动力装置及调控监测系统。目前,研究所拥有关于船舶的预研课题50多项。

　　改革开放以来,我所积极贯彻"科学技术为国民经济建设服务"和"科学技术是第一生产力"等方针政策,逐步转变观念,深化内部改革。1985年实行有偿合同制和经济自立试点,利用军工科研技术成果积极进行横向民用技术开发,使研究所逐渐适应社会主义市场经济发展形势,经济效益稳步发展。

　　回顾我所军转民的开发过程,有下面几点体会。

## 一、转变观念,狠抓科学管理,完善管理体制,向军民结合型（或科研经营型）研究所转变

　　我所建所的专业方向是从事舰用动力设备研制。20世纪80年代以来,在进行军品科研的同时,利用军工科研技术与设备,进行民品技术开发,在为国防建设服务的同时,为国民经济建设服务,使研究所由单一的军品研究所转变为军民结合型研究所。

　　研究所的这种转变,首先是单位领导和职工思想观念的转变。过去是封闭攻关出成果,闭门设计搞军品。人们的思想观念倾向于技术与经济分离,表现为重技术、轻经济;重成果、轻推广;工程技术人员只能搞研究设计,不能搞技术经营等。为此,我们在全所范围内提出了四个转变,即：由单一军品科研向军民结合转变;由单纯研究船舶动力向一专多能、多方位开发转变;由单纯的科研型向科研经营型转变;由自我封闭型向开放型、内外结合型转变。为实现这些转变,我所不断调整完善内部管理运行机制,尤其是结算办法和奖

---

* 文章发表时间：1995年4月；作者时任中国船舶重工集团公司第七〇三研究所所长。

励分配政策几乎是年年调整,起到了调动职工积极性和促进生产发展的作用。

## 二、以市场为导向,采取多种开发模式,创经济效益

我所民品开发始于1980年,以工业锅炉设计、制造率先进入市场。经过10多年的开发,已在石油、化工、冶金、电力、运输、机械、建材等行业取得了显著成绩,业务遍及20多个省、市、自治区。

研究所求生存、求发展的唯一出路是为市场服务;而市场的开发必须以科技为先导。充分发挥各专业技术优势,不断扩展市场是我所的立足之本。从20世纪80年代初期我所部分科技人员从事小规模民品技术开发开始,到1993年我所投标南山热电厂联合循环工程中标为止,历经了10多年的艰苦探索与拼搏,摸索出了如下的技术开发方式:

1. 以设计为主体,实行工程总承包。依据我所自身的技术实力和组织能力,以掌握的核心技术为中心环节,面向用户,实行方案设计、投标、技术谈判、产品设计、配套设备仪器购置、产品安装调试、产品试运行、岗位培训、技术操作、工艺规程制定等全过程的配套承包,即所谓"交钥匙"工程。

2. 面向社会,厂所结合。采用工厂和研究所结合的方式,把我所设计的产品和技术交给伙伴生产厂家,向其提供产品设计图纸、技术转让等。

3. 进入大型企业集团。为提高我所的承接能力和竞争能力,促进科研与生产相结合,我所联合一些大型企业集团公司,如大连船舶集团公司、东北热能电气集团公司、哈尔滨能源技术开发公司等。

4. 走向国际市场。开拓国际市场是我所的发展方向,如精密铸件出口日本、除氧器出口巴基斯坦、菱镁瓦自动生产线出口俄罗斯等。

5. 加速科技成果转化。依靠高新技术,把科技成果转化为商品,形成拳头产品或支柱项目,摆脱"找米下锅"的困难局面。

6. 疏导流通环节。使科、工、贸一体化,促进科技成果尽快转化为商品,转化为生产力。

由于积极开发,取得了较好的社会效益和经济效益,我所民品开发保持了稳步发展的势态。1988—1992年,我所每年经济收益保持在600万元左右;1993年成为研究所历史上经济形势最好年头,合同总收入6 200多万元,纯收益703.8万元,人均收入和人均收益也相应达到历史最高水平。

## 三、发挥军工专业技术优势,向民品开拓,直接为国民经济建设做贡献

如何走向社会?如何为国民经济建设服务?

答案是利用自己的专业优势。

我所的专业特点和技术优势是热能动力。几年来,我所就是围绕"能"字这个中心,向民品开拓。比如在燃气轮机专业方面,我所从燃气轮机电站的进口燃气轮机机组技术服务

做起,先后为国内8个电站(厂)从5个国家进口的7型燃气轮机机组服务。包括涉外谈判、清点设备、资料翻译、安装调试技术指导、运行保驾、培训操作人员等一条龙技术承包。随后,我所又在航空发动机改装,进口燃气轮机改造,地铁风机、压缩机几个方面进行发展,特别是在1991—1992年间,我所在深圳南山电厂完成了将3台PG5361P1单循环燃气轮机改造为回注蒸汽(双工质)热电联供的机组。该项目填补了一项国家空白,取得了历史性突破。几年来,我所逐步形成热电联供汽轮机、工业锅炉和余热锅炉、热能工程等15项民品"支柱",通过这些"支柱"带动全所民品开发工作迅速发展起来。

## 四、"敢"字当头,求实创新,攀登"高、难、新"技术,是研究所生存和发展的应备条件

近几年,我所民品开发向特区、向国家的重点工程、向中外合资项目倾斜。一句话,就是探索那些常人见而生畏的领域。

1989年,在国际投标中,我们战胜了一些国际的大公司,一举中标,承接了大亚湾核电站的3台预运行(启动)锅炉的研制项目。这个项目得来不易。众所周知,该工程的总经理是法国人,他从未见过中国什么地方有过这样大的启动锅炉,更不用说是由中国设计的。这个工程的资金,很大一部分是由国外厂商投资的。投标时,对这些投资的厂商是应该优惠的。另外像韩国的厂商,以现货供应和我所竞争,形势对我们并不利。但是,最终我们胜利了。

我们取胜的因素主要有两条:一是我们有技术储备,有信心、有决心完成此项工程;二是我们的报价仅为外商的1/3。经过近两年的努力,我们克服了技术上的许许多多困难,终于取得了该项目的成功。1992年元旦刚过,中央人民广播电台就播出了振奋人心的消息:第一台启动锅炉一次性点火试运行成功。后续其他两台也投运成功。

1990年,我们又利用"双工质"的军品预研成果,大胆地承接了深圳南山电厂的燃气轮机发电机组的改造项目。

在该项目的实施中,我们采用了先进技术改造旧装置,使之投入运行。这项技术成果填补了国内回注蒸汽循环燃气机装置自行研制和实际工程应用的空白,达到了世界先进水平。该项目可大幅度地提高电厂运行的经济性,又有利于环境保护,是一项符合我国能源政策、利国利民的高新技术和节能改造项目,必将在我国燃气轮机动力领域中产生深远影响。

1991年,我所承接了大庆石化总厂从美国引进的EC-301汽轮机的增容改造任务。这台机组的所有人都曾怀疑我们能否完成此项工程,主张请机组原设计厂家——美国的GE公司来完成。这台机组作为大庆乙烯生产的动力源,1985年投运,由于部件老化和磨损,以及原设计输出功率偏低,满足不了年产30万吨乙烯的需要。为此,大庆石化总厂才提出对该机组进行扩容改造。是我们行,还是老外行?美国GE公司和我所各自设计了该机组的改造方案。"不怕不识货,就怕货比货",两家方案摆出来,经联合评审组评审比出了高低,

我们的方案得到了认可。就连 GE 公司的代表也认为我所设计的方案,改造经费和周期均优于美国的方案,其中改造经费仅相当于美国方案的 1/4。

经过 10 个月的努力,我所完成了该机组的改造,并一次投运成功。1992 年 7 月 14 日投运至今,运行情况良好,输出功率由改造前的 20 790 kW 提高到 24 000 kW。改造前由于机组动力源不足,其乙烯生产从未达到过 30 万吨的设计生产能力;但经 1992 年下半年改造后,机组投运确保了动力源,当年产量就达到 32 万吨,为该厂增加了数亿元的产值。

1991 年,我所应胜利油田供电公司的请求,接替了原由德国承担的 V93 燃气轮发电机组的技术服务任务。

1991 年和 1992 年,我所先后承接了渤海油田的海上平台动力装置设备承包任务。该动力装置安装于海上平台上,要求总体体积小,相对集中,设备呈撬体结构。撬体分成换热器撬块和蒸发器撬块两部分。这样的结构在我们国内是少见的。经过投标竞争,我所中标。两套装置分别于 1992 年秋季和 1994 年 8 月投入运行,情况良好。

从以上所谈几个项目的完成情况中,我们充分认识到,只有"敢"字当头,敢于求实创新,敢于攀登高难新技术,研究所才能有生机,才能有发展。事实上,通过这些项目,我们完成了 5 500 多万元的产值,收益将近 2 000 万元,社会效益更巨大。同时我们还锻炼了技术队伍,特别是带起了一批年轻的工程技术人员,为今后的发展增添了力量。

# 双工质平行－复合循环发动机在船上应用的可能性*

闻雪友　邹积国

（哈尔滨船舶锅炉涡轮机研究所）

**摘　要**　双工质平行－复合循环发动机在工业电站已开始得到应用,其用作舰船主动力的前景如何? 文中从该循环的特点出发进行讨论,得出的结论是:这是一个富有吸引力的设想。

## 一、引言

现今,在舰船推进方面燃气轮机已成为一个重要角色。为此寻求船用燃气轮机的进一步发展和改进工作具有很大的吸引力。

在改进舰船燃气轮机循环性能方面,回热燃气轮机循环、中间冷却－回热燃气轮机循环和燃－蒸联合循环一直受到极大关注。然而当双工质平行－复合循环发动机(DFC)[1-3]问世之时,也应当引起一定的关注。

双工质平行－复合循环线图如图1所示。空气进入压气机1。2 为燃烧室。第二种工质(水)用泵5 送入热交换器6,从涡轮8 的排气(蒸汽与空气/燃料燃烧产物的混合物)废热中吸收热量变成蒸汽,然后进入燃烧室,用蒸汽来控制燃烧室出口温度。因此,在涡轮中是两种工质的混合物在膨胀做功,排气进入逆流热交换器6。从热交换器出口的混合气在开式系统中就排入大气,在闭式系统中则进入冷凝器9,凝水回收,气体排出。

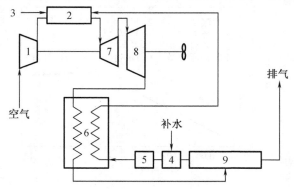

**图1　双工质平行－复合循环线图**

---

\* 文章发表时间:1987 年12 月。

## 二、采用双工质平行-复合循环发动机作为船舶主动力的优点

### 1. 技术政策的一致性

船用燃气轮机的发展与航空发动机的发展紧密相关,在整个发展过程中两者不可分割。航机改装可使造船部门以最快和最节省的方式应用先进的航机技术。这也是发展舰船燃气轮机的主要技术途径。有鉴于此,改进船用发动机循环的专门设计应该从其是否能基本上与"航空派生"这一政策相一致来加以考虑。双工质平行-复合循环在原则上是和用一个航空派生的燃气发生器的政策相一致的。当然,也可预料,核心发动机船用化的修改量将有所增加。

### 2. 高热效率,良好的变工况性能

在给定的涡轮进口温度下,循环效率、比功、蒸汽空气比(汽气比)和压比间的关系如图2所示。计算结果显示了采用双工质平行-复合循环后在效率上的增益。此外在同样的涡轮进口温度下,双工质循环最大效率的压比低于简单循环燃气轮机相应于最大效率的压比。为得到一个更实际、确切的比较,对一台9 500 kW的简单循环燃气轮机MGT-950做了改为DFC发动机的实际方案设计。发动机原循环效率为29.7%,改为DFC发动机后,相应于最佳效率的汽气比下,热效率达41%,如图3所示。

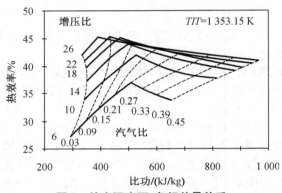

**图2  给定温度下,各相关量关系**

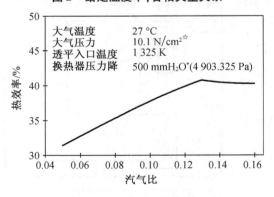

**图3  DFC发动机热效率**

---

☆1 N/m² = 1 Pa。

\* 1 mmH$_2$O = 9.806 65 Pa(准确值)。

双工质平行-复合循环发动机还具有良好的变工况性能。MGT–950 和 MGT–950DFC 的变工况性能如图 4 所示,DFC 发动机达到了低而平坦的油耗,因为其沿一实际的低能耗通路将两种工质带到高压、高温的运行条件。

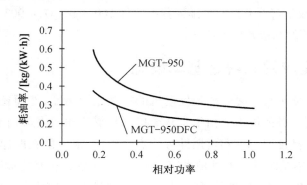

图 4　MGT–950 和 MGT–950DFC 的变工况性能

## 3. 高比功

由于第二工质为水,在等压下蒸汽比热容至少要比空气/燃料燃烧产物的比热容高一倍,这使双工质发动机达到高功率密度。

比功随汽气比增加而增加。从图 2 可见,对应于最大效率的汽气比也较大。因此,即使作为船用发动机,为减小尺寸质量,考虑在最佳效率点附近选取汽气比,也足以使发动机在达到高效率时有高比功。MGT–950DFC 相应于最大效率的比功(按压气机进口空气流量计算)较之原简单循环燃气轮机增加了 63%,其比功增大的能力令人印象深刻。显然,其单位功率的进气道面积较简单循环燃气轮机小得多。

可以注意到,MGT–950DFC 发动机的功率等级使其适用于典型的护卫舰、驱逐舰。

## 4. 最小的环境影响

DFC 发动机的排气温度相当低(见图 5),因而红外辐射特征小,这对舰艇有很大意义。排温的显著降低也使排烟道出口面积大幅减小,某种程度上补偿了余热锅炉的尺寸的劣势。

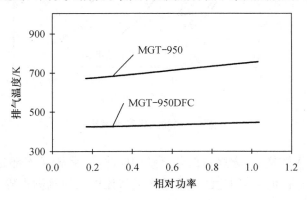

图 5　排气温度对比

喷注蒸汽的结果还改善了排放标准,使$NO_x$的含量大幅下降。

DFC发动机的性能对大气温度、排气背压以及燃气轮机部件效率等变化的敏感程度也均较简单循环燃气轮机小。例如,当大气温度从288 K提高到300 K时,对MGT-950DFC的参数,其循环效率仅下降0.6%。当排气背压增加500 $mmH_2O$(4 903.325 Pa)时,仅引起MGT-950DFC机1%的峰值热效率损失。

## 三、舰船可能应用系统之基本设想

以上已述及双工质平行-复合循环发动机的一些主要特点,然而还应同时保证不影响舰船的安全性和执行任务的能力。由于是从简单循环燃气轮机发展而来,应保留其固有的操作简单、启动迅速、高度可靠性和可用性等特点。因此,最可能在舰船上获得采用的是图1的基本系统。从基本系统发展的其他系统,虽然热效率或比功有所提高,但这是以系统的复杂程度、质量及尺寸增加为代价的,因而并不适宜在舰船上使用。

由于双工质平行-复合循环发动机在运行中需要高质量的水。为减小船上水耗,应考虑采用闭式系统。

1. 燃气轮机

对双工质平行-复合循环发动机来说,除了在燃气轮机循环设计中的两个独立参数(涡轮进口温度和压气机压比)外,尚有汽气比和热输入率(或空气燃料比)。关键在于适当选取循环参数或通过该部件间的独特匹配来达到高效率或高比功。对于一台拟改装为DFC发动机的实际燃气轮机来说,其涡轮进口温度、压比已基本确定,一旦对最大效率或高效率与高比功间折中的要求确定后,热输入率和汽气比也就相应确定了,如图6所示。对船用而言,为缩减余热锅炉尺寸,设计点应选择在接近最大效率处。

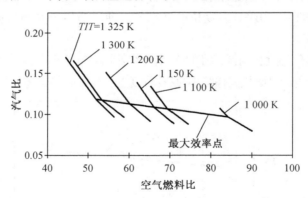

图6 热输入率和汽气比

无疑,因蒸汽喷入燃烧室,燃烧室部分要做相应改动,涡轮通流部分也需做相应的匹配调整。这意味着核心发动机船用化的修改工作量将比以往的航机改装更大。

2. 余热锅炉

喷注到燃气轮机中的过热蒸汽由余热锅炉产生,其为提高系统输出功率的关键部件,在相当程度上,锅炉尺寸就决定了系统内其他部件的尺寸。

在船用条件下选用直流锅炉具有明显的优点[4],与常规的汽鼓锅炉相比,其质量最轻、体积最小。由于直流锅炉的尺寸较小,其对负荷变化的响应更迅速,而且因为不存在汽鼓水位的控制、再循环控制等环节,使直流锅炉的控制更简单。直流锅炉接头少、简单、容水量小,而且水和蒸汽的全部容量也较小,增加了安全性。

双工质闭式系统的优点是对补给水的要求大为降低。但其对给水的水质要求高。双工质循环中蒸汽与燃气混合,使水质变差,这使水处理的问题变得突出。

由于余热锅炉入口的烟气温度较低,考虑把余热锅炉设计成允许"干烧"。这样当某些部件发生故障或要求进行非正常运行时,余热锅炉可干式运行。基础燃气轮机将不会因增加部分的故障而无法工作,从而保证舰艇不丧失动力。

3. 汽气冷凝器

为减小水耗,在闭式系统中采用冷凝器来冷却排气中的蒸汽,回收凝水。虽然这将增加压力损失和需要消耗辅助功率,但其影响小。

从热力学的观点来看,水容易回收。然而,事实上被冷却的工质是不凝气体和可凝气体的混合物,为了冷却排气混合气中的蒸汽,也必须同时冷却混合气中比例更大的不凝气。在冷凝器中,冷却和冷凝同时进行,传热、传质同时进行。与此同时,混合气体的组成、特性值和混合气的界膜导热系数均发生变化,其传热过程较复杂。在全部冷凝过程中,混合气中的水蒸气始终处于过热状态,即混合气体的温度始终高于混合气体中蒸汽分压力下的饱和温度。考虑到上述机理,要想实现蒸汽全部回收所需的换热面积是巨大的。但是,当考虑到燃料燃烧生成水以及空气中含水时,把产生蒸汽所加的水几乎全部回收又是可能的。

图 7 为 MGT–950DFC 机组的冷凝器中混合气的状态变化。图 8 为冷凝器中冷凝水量与混合气体温度间的关系。图 9 为冷凝器换热面积与冷凝水量间的关系。由图可见,所述的特定工作条件,使冷凝器的尺寸大幅增加(相对于相应的蒸汽轮机的冷凝器而言)。

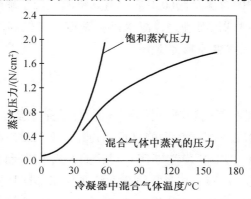

图 7　MGT–950DFC 机组的冷凝器中混合气的状态变化

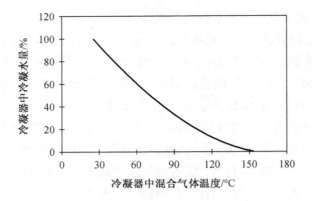

图 8　冷凝器中冷凝水量与混合气体温度间的关系

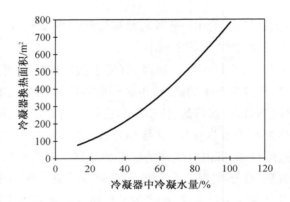

图 9　冷凝器换热面积与冷凝水量间的关系

DEC 装置的冷凝器还将遇到另一个问题：被冷却的介质中包含燃气，尽管用于舰船燃气轮机的海军专用轻柴油的含硫量低(0.2%)，但其壁温将低于硫酸的露点温度，应考虑腐蚀问题。设想采用钛管冷凝器，而且这样海水可在管道内以较高的流速通过，减少潜在污染或堵塞的危险，还将最大限度地减轻质量。

## 四、总体评估

对整个系统是否满足装船的各项主要要求进行评估是个复杂的问题。鉴于燃－蒸联合循环系统(COGAS)装置在船舶上应用的可行性讨论已久，加之 COGAS 与 DFC 在某些方面的类同性，因此，采用将 COGAS 装置与 DFC 对比的方式来进行讨论是适宜的。

燃－蒸联合循环如图 10 所示，仅是在概念上将布拉东和朗肯循环结成一体，两种工质有各自的环路，不发生混合。系统中除燃气轮机、余热锅炉、冷凝器和水处理装置外，还有蒸汽轮机和并体传动齿轮箱，因此 DFC 系统组成的简单性是显然的。

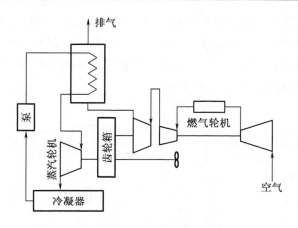

**图 10　COGAS 系统示意图**

两种循环中的余热锅炉是非常类同的,可以说具有基本相同的尺寸和质量。DFC 冷凝器的尺寸要比 COGAS 中蒸汽轮机冷凝器的尺寸大得多,但是在 DFC 系统中又省去了汽轮机和并体齿轮箱的空间和质量。比较的结果是,COGAS 在尺寸、质量方面略大于 DFC。

注意到 DFC 发动机的比功要比 COGAS 更大(如 MGT – 950DFC 的比功比 MGT – 950 COGAS 大 18%,两者在涡轮进口温度、热交换器换热面积及最小温差相等的条件下进行比较),因此 DFC 装置的质量功率比小于 COGAS,其单位功率的建造成本也低于 COGAS 装置。前已述及,DFC 装置单位功率的进气道以及出口烟道的面积也均较 COGAS 小。

综上,DFC 装置在船上布置的可能性至少与 COGAS 相同。

在经济性方面,虽然对于所讨论的 MGT – 950 的参数来说,改成 MGT – 950DFC 或 MGT – 950COGAS 后,两者在循环效率上的差异甚小。但是循环分析表明,随着参数提高,两者在效率上的差异将增大,如图 11 所示。

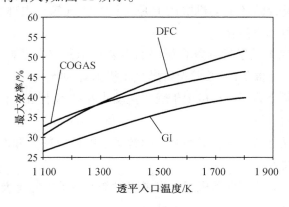

**图 11　循环效率对比**

最后,喷注蒸汽导致效率增加,从而使燃料成本下降,但能否足以抵偿水处理的成本? 通过对水/燃料成本计算分析,可直接确定所允许的最高处理水成本,在此成本费用下,处理水的成本正好与因效率增加而使燃料成本下降的值相抵偿。对 MGT-950DFC 在各种负荷下的计算结果列于表1。可以看出,按1986年的轻柴油价格 0.485 元/千克计算,处理水的价格可允许达 13.6~15.7 元/米$^3$。可以确定,DFC 发动机的燃料加水的运行消耗成本明显低于相应的简单循环燃气轮机的燃料成本。

表1 MGT-950DFC 各工况下运行成本

| 工况 | 耗油率/[千克·(千瓦时)$^{-1}$] | 汽气比 | 燃料成本/[元·(千瓦时)$^{-1}$] | 处理水成本/(元/米$^3$) |
|---|---|---|---|---|
| 100% | 0.285 | 0.000 0 | 0.138 2 | 0.00 |
|  | 0.210 | 0.131 5 | 0.101 9 | 14.25 |
| 80% | 0.300 | 0.000 0 | 0.145 5 | 0.00 |
|  | 0.222 | 0.117 5 | 0.107 7 | 15.72 |
| 62% | 0.320 | 0.000 0 | 0.155 2 | 0.00 |
|  | 0.240 | 0.112 5 | 0.116 4 | 13.78 |
|  | 0.335 | 0.000 0 | 0.162 5 | 0.00 |
|  | 0.250 | 0.110 0 | 0.121 3 | 13.56 |
|  | 0.365 | 0.000 0 | 0.177 0 | 0.00 |
|  | 0.260 | 0.107 5 | 0.126 1 | 15.21 |

## 五、结论

1. 在可预见的将来,舰船燃气轮机将继续走航机改装之路,燃用高质量燃料。双工质平行-复合循环在原则上是与这一特点相一致的。

2. 双工质平行-复合循环发动机实质上是直接在燃气轮机内部利用余热锅炉产生蒸汽,比 COGAS 的设计概念更简化。

3. 在舰船上采用双工质平行-复合循环的主要功用是节省燃料、增加功率,同时不影响舰船的安全性和执行任务的能力,并保留燃气轮机固有的特点。

4. 双工质平行-复合循环发动机用于舰船主动力在技术上是可行的,可望具有优于同功率 COGAS 装置的综合指标。

5. 双工质平行-复合循环发动机的突出问题是水的回收问题和对水质的要求高,后者依赖于先进的水处理技术。

6. 双工质平行-复合循环发动机用于未来舰船新的主动力是有吸引力的。

## 参 考 文 献

[1] CHENG D Y. Parallel-compound dual-fluid heat engine:US3978661[P]. 1976-9-7.
[2] CHENG D Y. Regencrative parallel compound dual-fluid heat engine:US412－8994[P]. 1978-12-12.
[3] CHENG D Y. regenerntive parallel compound dual-fluid heat engine:US4248039[P]. 1981-2-3.
[4] HALKOLA J T, CAMPELL A H, JUNG D. Racer conceptual design[J]. Journal of engineering for power, 1983, 105(3):621-626.

# 日本新一代船用燃气轮机(SMGT)研究计划[*]

闻雪友　(中国船舶重工集团公司第七〇三研究所)
　　　王明为　 (海军装备部)
刘培栋　(中国船舶重工集团公司第七〇三研究所)

**摘　要**:1997年,五家日本公司联手实施一项船用燃气轮机研究计划,目标是发展一种低$NO_x$排放(小于或等于1 g/(kW·h))、高热效率(38%~40%)的下一代船用燃气轮机。文中综述了该项目的目标、组织、计划、性能和关键技术。

## 一、引言

全球范围内的环境问题已成为公众关注的焦点问题之一,各国及当地政府也都在陆地区域采取了严格的排放标准。目前,在海事应用方面也已开始实实在在的排放控制。日本对近海船舶(包括渔船)的监测数据表明,在离海岸100 km范围内,其$NO_x$的排放量占日本国内排放总量的17%。1997年,国际海事组织正式通过一项议定书,该文件规定从2000年起对降低由船舶排放的$NO_x$必须有实际的计划。但是目前近海船舶几乎都用柴油机作推进动力,其热效率高,但$NO_x$排放量也很高。众所周知,燃气轮机的$NO_x$排放量远低于柴油机,因此在船舶上广泛装配燃气轮机可大大降低在海上排放的$NO_x$总量。超级船用燃气轮机(super marine gas turbine,SMGT)计划就是要发展一种下一代的船用燃气轮机,其既满足低$NO_x$排放,其热效率又能与当前的高速柴油机相匹敌。

## 二、组织

1997年,五家日本公司,即 Kawasaki Heavy Industries, Ishikawajima - Harima Heavy Industries, Daihatsu Diesel, Nigata Engineering 和 Yanmar Diesel,联合建立了超级船用燃气轮

---

[*]　文章发表时间:2003年1月。

机技术研究协会,开始实施超级船用燃气轮机计划。该项目获得 MOT,ASIS 和日本基金的支持,并被提作国家项目,按计划于 2003 年完成 2 500 kW 船用燃气轮机的原型机试验。

## 三、目标

2 500 kW 船用燃气轮机项目有以下三个特点:
(1) $NO_x$ 排放值低于 1 g/(kW·h);
(2) 热效率达 38% ~40%;
(3) 可用 A 型燃油(JIS K2205 No.1)。

该燃气轮机 $NO_x$ 的目标排放值大约仅为高速柴油机的 1/10,而热效率几乎与高速柴油机持平。与目前在用的工业燃气轮机比,其 $NO_x$ 排放降低了 1/3,热效率提高了约 10 个百分点(与同一功率等级的燃气轮机相比)。

## 四、计划

该研究计划于 1997 年起实施,周期为 6 年。第一年,重点进行新发动机的总体设计。在技术设计完成后将总的目标分解到各部件;1998 年开始对各部件进行严格的试验和分析,评估试验结果,其后进行施工设计;2001 年中期建成原型机;2002—2003 年进行整机试验,评估船用燃气轮机的性能。

## 五、SMGT 性能

表 1 列出了该发动机的目标性能。发动机的纵剖面图如图 1 所示。为使发动机热效率达到 38% ~40% 的目标值,SMGT 采用了回热循环,其循环计算结果如图 2 所示。如果采用简单循环,在燃气初温为 1 200 ℃时该功率等级的热效率上限为 35%。图 3 为船用燃气轮机的系统图。

表 1 SMGT 目标性能

| 性能 | | 单位 | 规格 | |
|---|---|---|---|---|
| | | | F 型 | V 型 |
| 额定功率 | | kW | 2 590 | 2 530 |
| 热效率 | | % | 39.1 | 38.4 |
| 空气流量 | | kg/s | 9.5 | |
| 压气机 | 额定转速 | r/min | 21 000 | |
| | 压比 | — | 8.0 | |
| | 等熵效率 | % | 84.0 | 83.7 |
| | 轴流级 | — | 4 | |
| | 离心级 | — | 1 | |
| 回热器 | 压力损失(空气侧) | % | 4.0 | |
| | 压力损失(燃气侧) | % | 4.0 | |
| | 回热度 | % | 83.0 | |
| | 形式 | — | 板翅式 | |
| 燃烧室 | 燃烧效率 | % | 99.0 | |
| | 压力损失 | % | 4.0 | |
| | 形式 | — | 4 管 | |
| 燃气发生器涡轮 | 额定转速 | r/min | 21 000 | |
| | 燃气初温 | ℃ | 1.200 | |
| | 等熵效率 | % | 87.5 | |
| | 轴流级 | — | 2 | |
| 动力涡轮 | 额定转速 | r/min | 13 000 | |
| | 等熵效率 | % | 90.2 | 89.2 |
| | 轴流级 | — | 2 | 3 |
| 排气管压损 | | % | 4.0 | |
| 燃气发生器机械效率 | | % | 98.0 | |
| 动力涡轮机械效率 | | % | 99.0 | |
| $NO_x$ 排放 | | g/(kW·h) | 1.0 | |

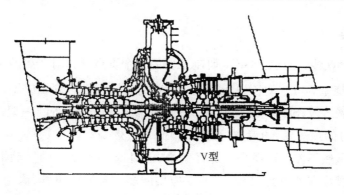

图 1 SMGT 纵剖面图

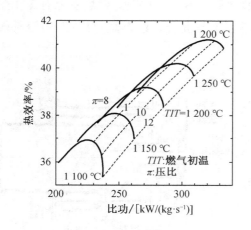

图 2 SMGT 循环计算结果

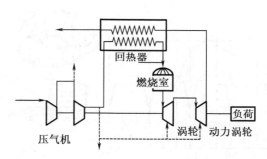

图 3 SMGT 系统图

## 六、压气机

压气机采用轴流加径流级的混合式结构以期达到 84% 的等熵效率。低压级(四级)为轴流,高压级(一级)为径流。轴流部分有两种方案:一种是 V 型,采用四级可调导叶,可使部分负荷下的效率更高;另一种是 F 型,仅进口导叶可调,可使额定功率下的效率最高。两种方案下轴流压气机的外径均为收缩状,在高效率下实现既定的压比是其目标,第一级是马赫数最高的地方,其最大相对马赫数也仅 0.9,以避免损失随马赫数增加而增大,第一级的温升为 31 K、第四级为 25 K。

为降低径流压气机的入口损失,进口气流有 20° 的预旋角,使相对马赫数降至 0.7。离心式叶轮有 24 个叶片,12 个长的、12 个短的,叶片出口的绝对马赫数为 0.8,出口扩压器不仅是要改进压力回收,而且还要降低外径,目前采用的是有 29 片叶片的扩压器,外径为 940 mm。

## 七、燃烧室

开发一种低 $NO_x$ 燃烧室是 SMGT 研制工作的一个重点。针对船用的特点,开发一种干式低 $NO_x$ 排放燃烧室为宜。对于以天然气为燃料的燃烧室已有许多应用,但燃液体燃料的燃烧室仍处于发展阶段,SMGT 的研究目标是通过试验和分析,研制一型适于用 A 型燃油(JIS K2205 No.1)的预蒸发、预混、贫油燃烧型低 $NO_x$ 燃烧室。

管式燃烧室虽然不如环形燃烧室那样紧凑,但对低 $NO_x$ 结构的发动机布置更为灵活,而且可方便地在燃烧室外进行维护,工业上应用颇多。若采用单管,涡轮入口温度周向不均度很大,因此不适宜用于高温,为此选择了四管,其具有较低的温度不均匀度,图 4 为两者的比较。图 5 为蒸发、预混、贫油燃烧型低 $NO_x$ 燃烧室的结构示意图。燃烧室上有三条燃油通路:一路到引燃室供点火和启动;一路到主燃室供预混合贫油燃烧;另一路引至补燃室。三者根据各种运行条件切换组合。

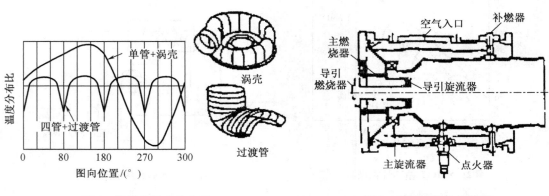

图 4　燃烧室温度分布图　　　　图 5　燃烧室结构示意图

## 八、涡轮

燃气发生器涡轮选用两级,可使涡轮均为亚音级(马赫数低于 0.8),叶片的扭角亦小(约 80°),可达高性能。若用 1 级跨音级,高气动负荷可能影响叶栅性能。由于燃气初温为 1 200 ℃,两级均采用气冷叶片。动力涡轮有两种方案:一种是 F 型,两级涡轮,设计目标是在额定工况下达到高效率;另一种是 V 型,三级涡轮,主要是为了改善部分负荷的热效率,这对降低经常变负荷运行的船用燃气轮机的运行费用是至关重要的,V 型的主要特征是动力涡轮 I 级导叶可调,在回热循环,配之以动力涡轮 I 级导叶可变几何,可提高部分负荷下的热效率。

然而,由于可转导叶机构位于高温区,需从压气机出口引气冷却运动部件;此外可转导叶片的顶部间隙有流动损失(在通常的涡轮结构中无此损失);再者在部分负荷下,I 级导叶变几何时,导叶的气动负荷增加,涡轮效率下降。为了防止在宽广的运行范围内有过大的气动负荷,在额定功率下仅给Ⅰ级导叶设定了一个小的气动负荷,使V型动力涡轮必须有三级。

V型和F型燃气轮机在部分负荷下的热效率比较见图6。

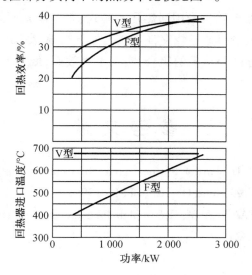

图6 部分负荷性能预测

## 九、回热器

众所周知,简单循环燃气轮机中燃烧产生的热能中有近70%随排气损失。因此,利用回热器回收废气中热能的回热循环将使燃气轮机具有更好的热效率。

对船用燃气轮机而言,必须设计一个紧凑的、高效的回热器。在回热度为83%的循环计算目标参数下,对四种形式的回热器进行了测试,结果表明:板翅式、一次表面式、管壳式和旋转式回热器的密封性差,仅适用于低压应用场合。虽然一次表面式回热器的尺寸比板翅式回热器更小,但最终决定采用板翅式,主要是由于板翅式回热器的强度、耐久性更好,且易于建造大尺寸的回热器。图7为板翅式回热器的尺寸计算结果,为达到83%回热度,大约需要2 m³的体积。

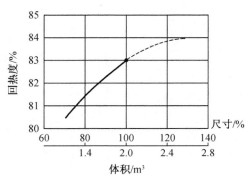

图7 板翅式回热器设计结果

## 十、结束语

日本新研制的船用燃气轮机采用回热开式循式(带紧凑式回热器)、混流式(轴流级+径流级)压气机,具有干式低排放燃烧室,1 200 ℃燃气初温,高温气冷涡轮以及压气机和动力涡轮部分级采用可转导叶等技术的组合,实现了高效率、低 $NO_x$ 排放和紧凑结构尺度的设计目标。

## 参 考 文 献

[1] SUGIMOTO T. R & D plan for the next-generation marine gas turbine (super marine gas turbine)[J]. IGTC99 Kobe OS-201,(11), 1999.

[2] KAWAMOTO M, TANAKA M. R&D plan for super marine gas turbine[J]. Bulletin of GTSJ, 2000, 2000:27-29.

# 国产航空斯贝发动机的工业及船用化发展*

闻雪友　赵友生

(哈尔滨船舶锅炉涡轮机研究所)

**摘　要**：文中介绍了改型航空发动机应用概况，提出了将航空斯贝发动机作为母型机改装为工业和船用系列化机型的途径及优点。

## 一、引言

国际上，工业和船用燃气轮机正方兴未艾。在国内，相关行业对工业及船用燃气轮机的热情一直在增长，我国各大油田相继从国外购置了各种型号的燃气轮机发电、热电或热动联供装置(表1)，由国产航空发动机改装的轻型燃气轮机也纷纷投运，用于发电、供热、注水、泵气、煤矿灭火和机车动力等(表2)。在舰船动力方面，除了国内研制的船用燃气轮机外，从美国购置了LM2500燃气轮机，以及正在和美国普朗特·惠特尼公司合作开发的FT-8工业及船用燃气轮机，也显得生机勃勃(表3)。

表1　船用燃气轮机发动机

| 序号 | 机组型号 | 厂商名称 | 投运方式 |
|---|---|---|---|
| 1 | MS6001 | GE | COGAS发电,热电联供 |
| 2 | Frame6 | JB | COGAS发电,热电联供 |
|   |   | Alshom | COGAS发电 |
| 3 | SK15HE | R—R | 热电联供 |
| 4 | Centaur | Solar | 海洋平台发电,驱动压气机 |
| 5 | Tornado | Ruston | 热动(驱动压气机)联供 |
| 6 | KG5 | Kongsberg | 热电联供 |
| 7 | LM2500 | GE | 发电 |
| 8 | PG5000 | Ruston | 热动(驱动压气机)联供 |

---

\* 文章发表时间：1988年7月。

**表2 航空改型燃气轮机应用概况**

| 型　号 | 燃气轮机 | 功　率/kW | 应用地点 |
|---|---|---|---|
| YD-2000 发电机组 | $WJ_6G1/1A$ | 2 000 | 中原、南阳、任丘、长庆、克拉玛依等油田 |
| YD-1250 发电机组 | $WJ_5G$ | 1 250 | 大庆油田 |
| $3DR_5$ 热电联供电机组 | $WP_6G1/1A$ | 4 050/4 700 | 大庆油田 |
| WB-75 注水泵机组 | $WZ_6G$ | 750 | 中原油田 |
| WZ5G 注水煤机组 | $WZ_5G$ | 1 140 | 炼油厂,克拉玛依油田 |
| WZ5G 压缩机组 | $WZ_5G$ | 1 140 | 中原油田 |
| DQ-1000 煤矿灭火器 | $WZ_5G$ | — | 在萍乡煤矿作试验 |
| $WS_9G_1$ 发电机组 | $WS_9G1$ | 4 400 | 东方红炼油厂 |
| 机车动力 | $WJ_6G_2C$ | 2 600 | 试运行 |

**表3 国内船用燃气轮机近况**

| 型　号 | 产　地 |
|---|---|
| LM2500 | 美　国 |
| MGT-260 | 中　国 |
| MGT-950 | 中　国 |
| FT-8 | 中美联合 |

我国于1975年从英国罗尔斯-罗伊斯公司引进的Spey MK202航空涡轮风扇发动机已基本国产化,其工业及船用化工作取得了进展。目前,正是讨论其进一步发展的适当时机。

## 二、斯贝——合适的母型机

航空斯贝发动机有民用和军用两个系列,其发展系列如表4所示、主要性能数据如表5所示、发展过程如表6所示。

从中可见,航空斯贝发动机经各种机型的发展和改进,性能有了很大的提高。燃气初温由1 313 K提高到1 440 K,总压比由15.4增高到21.4,内涵流量由46 kg/s加大到56~68 kg/s。到1981年底,其使用总数已达4 440台,累计运行$2.2 \times 10^7$ h,发展得相当成熟,因而该发动机是一种适于工业、船用化改装,性能良好且功率等级适中的机型。

**表 4　航空斯贝发动机的发展系列**

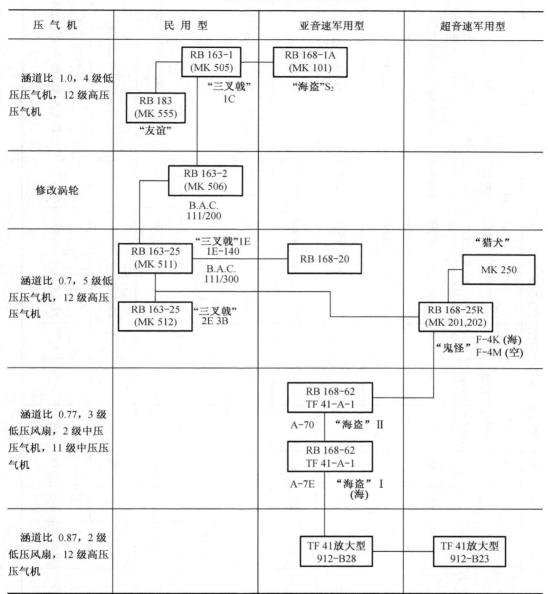

表 5　航空斯贝发动机的主要性能数据

| 发动机型号 | 低压压气机 ||| 高压压气机 ||||| 涵道比 | 总比压 | 涡轮前温度/K | 最大起飞推力/kg | 最大连续推力/kg | 备注 |
|---|---|---|---|---|---|---|---|---|---|---|---|---|---|---|
| | 级数 | 压比 | 流量/(kg/s) | 转数/(r/min) | 级数 | 压比 | 流量/(kg/s) | 转数/(r/min) | | | | | | |
| RB183 (MK555) | 4 | 2.46 | 92 | 8 600 | 12 | 6.26 | 46 | 12 200 | 1.0 | 15.4 | | 4 468 | 4 295 | 军用运输机,1368年服役 |
| RB163-1 (MK505) | 4 | — | 92 | — | 12 | — | — | — | 1.0 | 16.9 | 1313 | 4 468 | 4 285 | 民航机 |
| RB163-2 (MK506) | 5 | — | 92 | — | 12 | — | — | — | 1.0 | 16.9 | 1 313 | 4 722 | 4 532 | 民航机 |
| RB163-26 (MK510) | 5 | — | 93.3 | — | 12 | — | — | — | 0.7 | 19.1 | 1 356 | 4 990 | 4 781 | 民航机 |
| RB163-25 (MK511) | 5 | — | 93.3 | — | 12 | — | — | — | 0.7 | 19.1 | 1 356 | 5 171 | 4 963 | 民航机,1968年航线使用 |
| RB163-25 (MK512) | 5 | 2.6 | 93.3 | 8 115 | 12 | 7.35 | 55 | 12 490 | 0.7 | 19.1 | 1 360 | 5 430 | 5 253 | 民航机 |
| RB165-25R (MK201-202) | 5 | 2.7 | 95 | 8 600 | 12 | 7.45 | 56 | 12 640 | 0.7 | 20.1 | 1 440 | 5 670 | — | 攻击机,1968年交付使用 |
| RB168-62 (TF41-A-1) | 3/2 | 2.45/1.39 | 117 | 8 850 | 11 | 5.92 | 66 | 12 770 | 0.77 | 20.2 | — | — | — | 攻击机,1968年交付使用 |
| RB168-66 (T41-A-2) | 3/2 | 2.49/1.39 | 119 | 9 150 | 11 | 8.18 | 68 | 13 000 | 0.77 | 21.4 | 1424 | 6 800 | — | 舰载攻击机1968年交付使用 |

表6 航空斯贝发展过程

| 机　型 | 发展过程 $\left(\dfrac{\text{型号}}{\text{千克推力}} \times \text{推力增长率}\right)$ | 备　注 |
|---|---|---|
| 斯贝（民用） | $\dfrac{\text{MK505}}{4\,720} \times 9.5\% \to \dfrac{\text{MK511}}{5\,170} \times 5.5\% \to \dfrac{\text{MK512-5}}{5\,454} \times 4\% \to \dfrac{\text{MK512-4}}{5\,670}$ | 共增加 20.1% |
| 斯贝（军用） | $\dfrac{\text{MK101}}{5\,100} \times 6.6\% \to \dfrac{\text{MK250}}{5\,440} \times 4.6\% \to \dfrac{\text{MK201}}{5\,692} \times 12.4\% \to \dfrac{\text{TF41-A1}}{6\,400} \times 6.2\% \to \dfrac{\text{TF41-A2}}{6\,800}$ | 共增加 33.3% |

美国 COOPER-ROLLS 公司与英国 ROLLS-ROYCE 公司已先后将航空斯贝 RB168-66型(TF41-A2)发展成五型工业及船用燃气轮机。

(1) COOPER-ROLLS 工业斯贝燃气轮机

由 MK1900 工业斯贝燃气发生器(低压风扇顶切)配 RT45 动力涡轮组成,用于加拿大气干线。

(2) SMIA 船用燃气轮机

由 MK1903 船用斯贝燃气发生器(低压压气机系重新设计并采用 RAB 燃烧室)配 1978 动力涡轮组成,已用作英国海军 22 型、23 型和荷兰海军 M 级等大型护卫舰的主动力。

(3) SK15HE 燃气轮机热电联供装置

其燃气轮机系由 MK1907 工业斯贝燃气发生器(除燃烧室及启动方式外与 MK1903 相同)配侧面排气的 1978 动力涡轮组成,用于我国大庆、南疆油田等处。

(4) SMIC 燃气轮机

1984 年起,R-R 公司在 SMIA 的基础上通过改进高压涡轮叶片冷却结构提高了燃气初温,加大低压压气机压比和流量,使功率增大到 18 MW,发展成 SMIC 燃气轮机。

(5) SMIC ICR 燃气轮机

在 SMIC 基础上增加一个中间冷却器和一个回热器,发展成间冷-回热循环。其额定功率可达 21.5 MW,热效率达到 42%~43%,具有极好的部分负荷性能。这种发动机正在研制中。

以上情况说明,斯贝是一台合适的母型机。

英国在十多种斯贝型号中选择了 RB168-66(TF41-A-2)作为工业/船用改装的母型机,主要有三种原因:

(1) 其核心机流量大,改装后功率大;

(2) 为美国海军 A7E "海盗" Ⅱ 歼击机设计,有关部件已考虑了抗腐蚀要求,因此特别适合于船用改装;

(3) 风扇级数少(两级),有利于实施顶切方案。

我国引进的是 MK202,其内涵流量比前者小 17.6%,若燃气初温相同,则功率也明显小于前者。MK202 有五级风扇,对于顶切方案,难度增加。

## 三、中国工业和船用斯贝的发展系列

目前,我国工业和船用燃气轮机方面还处于较落后状态,研制经费严重不足,因此需要最大限度地一机多用。此外,双转子内、外涵涡轮风扇发动机斯贝的涵道比小,推力中等,发展成熟,客观上也有可能发展成各种功率等级、性能的工业及船用燃气轮机,以满足各种应用场合的需要。

图1为Spey MK202发动机组成示意图。

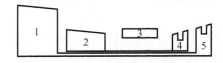

**图1　Spey Mk202 发动机组成示意图**
1—低压风扇;2—高压压气机;3—燃烧室;4—高压涡轮;5—低压涡轮

(1)利用MK202核心发动机发展一型工业/船用双轴燃气轮机。其燃气发生器由原高压压气机、燃烧室和高压涡轮组成,功率由新设计的动力涡轮输出(图2)。此型燃气轮机功率可达4 420 kW,耗油率350 g/(kW·h)。

(2)上述简单循环燃气轮机压气机的压比为8,可在此基础上发展成回热循环(图3),从而使热效率达35%。当然,一台高压比的简单循环燃气轮机也可能达到这样的效率水平,且免去了增加回热器所带来的复杂性。两相比较,是否一定要用回热循环值得商榷。当然,如果没有一台适宜的高压比简单循环燃气轮机,则发展回热循环也不失为一种现实的考虑。

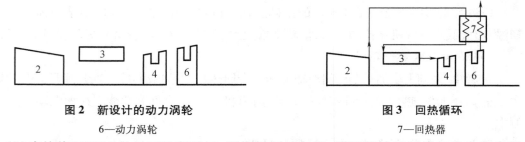

图2　新设计的动力涡轮　　　　　　　图3　回热循环
6—动力涡轮　　　　　　　　　　　　　7—回热器

(3)直接将MK202的低压风扇顶切,用作低压压气机,利用双转子的燃气发生器发展一型中档功率的工业/船用三轴简单循环燃气轮机(图4)。功率为10 500 kW,耗油率为290 g/(kW·h)。

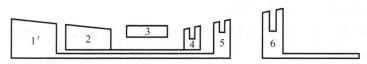

**图4　中档功率工业/船用三轴简单循环燃气轮机**
1′—低压压气机

(4)在上述低压风扇顶切形成的低压压气机前再加一级,提高压气机的流量和压比,使功率增至 12 000 kW,发展成一型工业用燃气轮机。

（5）把 MK202 双转子燃气发生器中的低压风扇整个去除,换装一台重新设计的低压压气机（压比 3.3）,使整机功率达 12 500 kW。这一型与前述的"顶初加级"型相比不仅功率略大,并可获得较好的低工况性能,因而更适于发展为一型工业/船用燃气轮机。

如果在新设计的低压压气机前再加一级,进一步增大流量、压比,功率可达 14 000 kW。这个方案的最大优点是原航机的低压涡轮几乎可不作调整,高压轴转速也可保持不变,给燃气发生器的匹配调整带来许多方便。

（6）在方案（5）的基础上增加中间冷却器和回热器,构成间冷-回热复杂循环燃气轮机（图 5）。其具有很高的循环效率、优良的变工况性能。由于热交换器方面的技术进展,这一方案仍能使装置结构紧凑。这种燃气轮机可能特别适于舰用,美国、英国和联邦德国都在积极研制,作为未来的舰用燃气轮机动力。

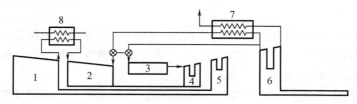

图 5　间冷-回热复杂循环燃气轮机

8—中间冷却器

图 6 示出了英国 ICR Spey 的功率、耗油率关系。与其他循环相比较,其热效率可达 42%。利用动力涡轮可变几何,使该型机在 50% 的功率时仍能保持全负荷的效率。

然而,ICR 循环的研制更为复杂。

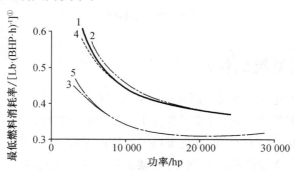

图 6　斯贝改型机功率与耗油率的关系

1. SMIA/MK1907（双转子）;2. SMIC（双转子）;3. ICR（双转子、带中间冷却和回热器）
4. 单转子简单循环;5. 单转子、带回热器

---

① 1 lb = 0.453 592 37 kg（准确值）, 1 BHP = 0.746 kW。

(7) 可在斯贝改型简单循环燃气轮机的基础上发展成双工质热机(图7)。把余热锅炉产生的蒸汽回注到燃气轮机中,燃气-蒸汽混合工质在涡轮中做功。这种循环的热机比功大、热效率高。以方案(3)的 10 500 kW 燃气轮机为例,改成双工质发动机后,其功率可增至 15 000 kW,热效率提高 38%(以纯发电计)。这种双工质热机特别适合于工业上热电联供,当余热锅炉前有补燃时更具有宽广的运行范围,可以相当灵活地适应各种热电负荷要求。

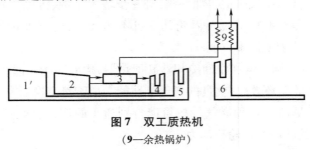

图 7　双工质热机
(9—余热锅炉)

(8) 在斯贝改型简单循环燃气轮机上加装余热锅炉(图8),产生的蒸汽驱动-蒸汽轮机,组成燃-蒸联合循环装置。以方案(3)的 10 500 kW 燃气轮机为例,改为联合循环后,其总功率为 15 000 kW,热效率为 43%。

如果所设计的余热锅炉产出的蒸汽系供热之用,则组成燃气轮机热电联供或热动联供装置。

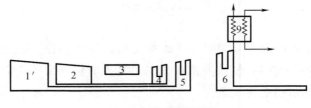

图 8　加装余热锅炉

## 四、系列机型发展的优点

以 MK202 为基础发展的系列机型具有显著的优点:
(1) 用一种灵活的组合概念,适应了各种不同的用途(图9);
(2) 为实现系列化、通用化和标准化提供了良好的基础;
(3) 系列发展方式无疑能降低研制费用,减少用户的购置成本;
(4) 便于解决用户最关心的售后服务保障问题;
(5) 易于积累运行经验,提高设备可靠性;
(6) 有利于培养操作人员。

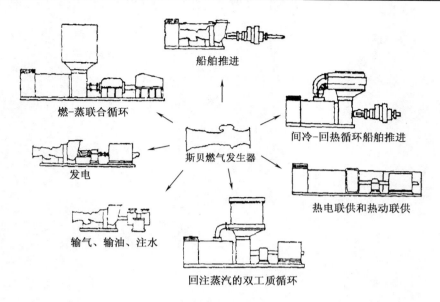

图 9　国产斯贝改型机的应用

## 五、结束语

(1) 国产航空斯贝发动机是一台合适的母型机。其工业/船用系列化发展是一项符合国情,技术上可行,有一定先进性、现实性的政策,希望引起各有关部门进一步重视。

(2) 在发展工业/船用系列过程中,应对能燃烧多种液态和气态燃料的能力以及控制装置予以足够注意。

(3) 系列发展问题并非单纯设想。事实上,双轴简单循环燃气轮机,三轴简单循环燃气轮机,顶切加级低压压气机,重新设计低压压气机,双工质平行-复合循环验证机等方案分别在哈尔滨船舶锅炉涡轮机研究所、西安航空发动机公司、中科院工程热物理所、哈尔滨船舶工程学院等单位进行了大量实际性工作,有的已开始整机运行。

# 二、蒸汽回注燃气轮机技术

# M70 系列燃气轮机的发展与应用*

闻雪友

(中国船舶重工集团公司第七〇三研究所)

**摘 要**：介绍了 M70 系列燃气轮机的发展模式和途径，以及在气垫船上的运行经验；综述了其在气垫船和排水型水面舰船上的实际应用。

## 一、引言

乌克兰的"曙光－机器设计"科研生产联合体是一家船舶工业燃气轮机的设计、研究和生产企业。至今，该企业设计制造的发动机和机组已超过 3 300 台，其产品用于舰船、天然气增压站、电站等。

基于对"曙光－机器设计"企业和"大海"船厂等的考察，文中专题介绍了由"曙光－机器设计"企业研制的 M70 系列船用燃气轮机的发展及其在舰船上的应用情况。

## 二、M70 系列燃气轮机的发展过程

M70 燃气轮机是"曙光－机器设计"科研生产企业按苏联海军订货协议研制的第三代船用燃气轮机的第一个基础发动机。

M70 燃气轮机从开始就有两个设计方案：

（1）前输出、不可倒车的发动机，用于气垫船；

（2）后输出、可倒车的发动机，用于排水型水面舰船。

近几年，又发展了前后同时输出、不可倒车的ДП73 发动机，其用于气垫船。

在 M70 燃气轮机发展过程中，不断地进行各种改进研究，包括对气动性能、结构强度、燃气初温、材料涂层和系统的研究。针对各项研究制订的研究计划的完整性和质量检查，由苏联海军科学研究所（НИИВМФ）和造船工业部（МСП）科学研究所共同进行。

---

\* 文章发表时间：2006 年 8 月。

从1970年开始研发至今,气垫船用与排水型船用M70燃气轮机分别经历了12次和6次改进,所有设计、试验研究工作依照与海军及其他用户签订的16个合同依次进行。在此过程中,M70燃气轮机的寿命从100 h提高到了20 000 h;每个研发阶段均由各部门联合组成的委员会进行试验考核;对4台样机进行了当量寿命试验。

据初步统计,至1995年底,M70系列燃气轮机的应用情况见表1。

**表1 M70系列燃气轮机的运行情况**

| 舰船类型 | 气垫船 | | 增压站 | 反潜舰 | $T_3 < 1\ 100\ ℃$ | | 护卫舰 | 水翼客船 | $T_3 = 1\ 132\ ℃$ | |
|---|---|---|---|---|---|---|---|---|---|---|
| | | | | | 水翼护卫舰 | 导弹巡洋舰 | | | 导弹快艇 | 水翼海岸艇 |
| 燃气轮机型号 | ДК71<br>ДМ71<br>ДР71 | ДП71<br>ДП79*<br>ДП73* | ДТ71 | ДС71 | ДС71 | ДН71<br>ДС71 | ДС71 | МО37 | ДР77<br>ДС77<br>ДО77 | ДС77 |
| 功率/kW 正车 | 4 413/<br>7 355 | 7 355 | 8 311 | 6 620 | 7 355 | 5 663 | 6 546 | 7 355 | 7 355/<br>8 826 | 7 355 |
| 功率/kW 倒车 | | | | 1 103 | 1 103 | 1 471 | 2 921 | 735 | 735 | 735 |
| 燃机数量/台 | 66 | 45 | 14 | 16 | 2 | 6 | 2 | 2 | 144 | 24 |
| 总运行时间/h | 27 500 | 15 700 | 85 000 | 30 000 | 1 000 | 25 000 | 7 000 | 11 600 | 53 000 | 20 000 |
| 运行时间最长的发动机工作时间/h | 600 | 920 | 13 250 | 2 750 | 660 | 5 030 | 3 810 | 10 000 | 2 050 | 1 000 |

注:*为1995年后发展的型号,数量未计入。

从表1中可知:在气垫船上已有111台燃机应用,累计运行约43 241 h;在排水型舰船及客船上已有196台燃机应用,累计运行约147 600 h;在天然气压缩站已有14台燃机应用,累计运行85 000 h;总装机台数321台,总累计运行时间约275 800 h;单台发动机最长的工作时间为13 250 h。可见,该机型在长期研发过程中已日趋完善、成熟。

ДИ70(ДН70),即GT-10000燃气轮机是"机器设计"科研生产企业于1996年在ДП71船用燃气轮机的基础上研制的,已用于驱动天然气管线压缩机,是新型工业用燃气轮机。

发展过程中形成的有代表性的燃气轮机型号(新编号)有GT6000,GT6000+与GT10000,其主要性能见表2。

表2  GT6000、GT6000+和GT10000燃气轮机的主要性能

| 燃气轮机 | GT6000 | GT6000+ | GT10000 |
|---|---|---|---|
| ISO条件下的功率/kW | 6 700 | 8 300 | 10 100 |
| ISO条件下的效率/% | 31.5 | 33.0 | 36.0 |
| 压气机压比 | 13.9 | 15.7 | 19.5 |
| 排气流量/(kg/s) | 31.0 | 33.4 | 37.2 |
| 排气温度/℃ | 420 | 442 | 458 |
| 动力涡轮转速/(r/min) | 7 000 | 7 000 | 6 500 |
| 燃气轮机质量(不含底架)/kg | 2 900 | 3 000 | 4 000 |

GT6000+(ДП71)燃气轮机是专门为"祖伯尔"(ЭУБР)气垫登陆艇研制的。

"祖伯尔"气垫船的动力装置由3套燃气轮机驱动的空气螺旋桨和2套燃气轮机驱动的垫升风扇组成。燃气轮机从气垫舱内吸气,设计点的进气温度较高,并要求在规定船用条件下达到较高的出力。在GT6000燃气轮机的基础上发展而来的GT6000+燃气轮机采用的主要技术途径是提高燃气初温,并增加燃气发生器高、低涡轮转子转速。与GT6000燃气轮机相比,GT6000+燃气轮机的压比增加了13%,流量增加了7.7%,功率增加了24%,效率提高了4.8%。

为保证参数提高后燃气轮机工作的可靠性,对GT6000燃气轮机采取了如下主要措施:

(1)高、低压涡轮的通流部分采用更好的高温耐热合金材料和抗蚀涂层;

(2)高压涡轮导叶采用双支承结构,涡轮第1~4级盘均采用空气冷却;

(3)高、低压涡轮采用单独支承。

从表2中可见,与GT6000+燃气轮机相比,GT10000燃气轮机的功率增加28%,效率提高9%。其所采用的技术途径如下:

(1)在M70燃气轮机低压压气机前,增加了带可转进口导叶的低压压气机"零"级,使流量增大约11.4%,压比增加约24%;

(2)重新设计了高、低压涡轮的导叶,在高压涡轮动叶上方增加可调径向间隙,并采用新的高压涡轮轮盘;

(3)重新设计了动力涡轮。

## 三、装用M70系列燃气轮机的气垫船

苏联时期,气垫船的主要设计单位是"金刚石"设计局。

1. "卡里玛尔"

装用M70燃气轮机的第1艘气垫登陆艇"卡里玛尔"(КАЛЬМАР)的动力装置由2套MT70K机组组成,每套机组由1台功率为7 355 kW的ДК71燃气轮机(耗油率258 g/(kW·h),寿命500 h)以及将功率分别分至垫升风机和推进空气变距桨的传动装置组成。首艇于1972年

试航,后来在"大海"船厂批生产,建造了22艘。ДК71燃气轮机累计运行20 508 h。

从第1艘试验艇和批生产的首艇上均明显地看出,M70燃气轮机需在以下方面进行改进:

(1)针对进气除水装置不良的改进。在机舱和发动机通流部分落下大量飞溅的海水珠,尤其是在艇以很低的速度垫升和登滩的工况下,需减少随空气进入发动机通流部分的舷外水量。

为此,选择最佳的进气口转换时间,把从大气吸气转至从集气室吸气;反之亦然。

(2)针对M70机组水平传动装置振动大的改进。采取的措施是精确地对接弹性轴、减速器法兰和中间支承。

(3)针对发动机通流部分积垢的改进,在发动机启动和加速时压气机表现不良。改进措施为:在每次出海后用移动式清洗装置清洗,并完善燃气轮机通流部分清洗工艺。

2. "螯虾"

1978—1980年,"大海"船厂建造了2艘"螯虾"(OMAP)气垫登陆艇,其动力装置为1套M34机组,由功率为4 413 kW(6 000 hp)的ДМ71燃气轮机组成,累计运行1 162 h。

在试验中曾发现该型首艇减速传动装置偏心,原因是艇中翼的刚性不足,而减速传动装置正是安装在中翼上。后增强了中翼结构,问题随之解决。

3. "燕"

1983年,"大海"船厂建造了2艘"燕"(KACATKA)级气垫登陆艇,在黑海运行,其动力装置为2套МТ70М机组和2台ГТТ–100燃气轮机发电机组。МТ70М机组由功率为7 355 kW(10 000 hp)的ДМ71燃气轮机和减速传动装置组成;ГТТ–100燃气轮机机组的功率为100 kW。ДМ71燃气轮机总计运行1 221 h。

4. "海鳝"

哈巴洛夫斯基造船厂在20世纪80年代曾建造8艘"海鳝"(MYPEHA)级气垫船,每艘船装2套МТ70Р机组和2台ГТТ–100К燃气轮机发电机组(每台功率为100 kW)。每套МТ70Р机组由1台功率为7 355 kW的ДР71燃气轮机和将功率分至垫升风扇和空气推进螺旋桨的减速传动装置组成。该型船从1985年开始运行,ДР71燃气轮机总计运行4 667 h。

5. "祖伯尔"

通过积累上述气垫船的设计、制造和运行经验,"金刚石"设计局和滨海船厂设计和建造了中型气垫登陆艇"祖伯尔"。该艇上装有2套M35–1增压机组(每套由功率7 355 kW,耗油率258 g/(kW·h),寿命为1 000 h的ДП71燃气轮机和驱动空气调距桨的传动装置组成)以及4台ГТТ–100К燃气轮机发电装置(每台功率为100 kW)。

首艇进行了一年半的结构试验后,于1988年交苏联海军进行试验运行。迄今,"祖伯尔"登陆艇在滨海船厂建造了5艘,在"大海"船厂建造了4艘,其ДП71燃气轮机总计运行15 683 h(首台运行915 h),ГТТ–100К燃气轮机运行了22 998 h(首台运行2 003 h)。

在对首艇进行结构试验和运行试验期间,以及在批生产艇运行过程中,均发现在推力发动机通流部分有外物引起的损伤。外物是从艇柱墩内的空气道吸入发动机的,柱墩内表面的刚性肋骨处聚集了异物,在发动机准备启动和工作时无法有效清理。

为此,用专门的光滑覆面材料封闭了柱墩内部的刚性肋骨,在发动机进口装设了防护网,并采取了防止异物在空气道中聚集的限制措施。运行结果证明了上述措施非常有效。

另外,还发现水平齿轮传动装置中风扇减速器的基础刚性不足,要求在新建和修理的艇上增加其刚性。

M70系列燃气轮机装备气垫船的情况汇总见表3(截至1995年底),相应的动力装置示意图如图1所示。

表3 M70船用燃气轮机发动机装备气垫船情况汇总

| 气垫船名称 | 数量/艘 | 燃机数量/台 | 燃机累计运行时间/h |
|---|---|---|---|
| "卡里玛尔" | 22 | 44 | 20 508 |
| "鳌虾" | 2 | 2 | 1 162 |
| "燕" | 2 | 4 | 1 221 |
| "海鳝" | 8 | 16 | 4 667 |
| "祖伯尔" | 9 | 45 | 15 683 |
| 合计 | 43 | 111 | 43 241 |

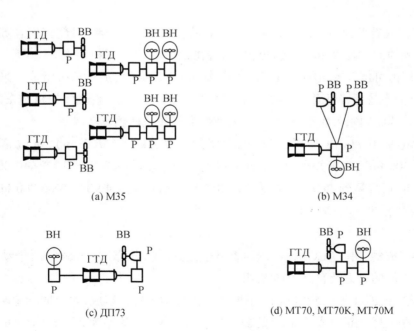

图1 气垫船动力装置示意图

## 四、装用 M70 系列燃气轮机的水面舰船

1."天蝎座"

1979 年,"天蝎座"(AHTAPEC)级水翼海岸警备艇交付苏联海军舰队。

该艇的动力装置 M20 机组由可倒车的 ДC77 燃气轮机和角传动减速器组成。ДC77 燃气轮机功率为 7 355 kW,耗油率为 285 g/(kW·h),倒车功率为 735 kW,翻修寿命为 1 000 h。

2."闪电"

1980 年,"闪电"(МОЛНИЯ)级导弹快艇首艇完成海试服役。该艇的动力装置 M15 由巡航机组和加速机组组成。巡航机组与加速机组间可以联合工作,也可单独工作。

巡航机组由两台可倒车的小功率燃气轮机和两台巡航行星双速齿轮箱组成。两台巡航减速器间用跨接轴连接,可以将任一台巡航机组的功率分至两根螺旋桨轴,以保证艇经济的航行;加速机组由两台可倒车的ДО77(ДС77)燃气轮机和两台角传动齿轮箱组成。ДО77 燃气轮机的功率为 8 826 kW(12 000 hp),耗油率为 285 g/(kW·h),空气流量为 34 kg/s,压比为 16,燃气初温为 1 110 ℃,寿命为 1 000 h。

后来建造的"闪电"级艇中,也有采用柴-燃动力方式的,每舷均用功率为 2 942 kW(4 000 hp)的柴油机代替巡航燃气轮机。

3."亚特兰大"

1982 年,"亚特兰大"(АТЛАНТ)级巡洋舰的首舰——"光荣"号舰完成在苏联国家验收试验。该舰的动力装置 M21 机组也包括加速机组和巡航装置机组。

巡航机组由可倒车的 ДН71(ДС71)燃气轮机、余热锅炉、蒸汽轮机组成。采用燃-蒸联合循环,耗油率较低,从而使舰船能保持规定的续航力。当巡航燃气轮机倒车时,蒸汽轮机自动断开。苏联国家验收委员会指出,其巡航燃气轮机的倒车功率不足。

巡航机组的 ДН71(ДС71)燃气轮机功率为 5 663 kW(7 700 hp),蒸汽轮机功率为 1 692 kW(2 300 hp),二者并车输出。巡航机组(1 轴)总功率为 7 355 kW(10 000 hp),寿命为 10 000 h(后发展到 ДС71,寿命为 20 000 h),续航功率为 4 928 kW(6 700 hp),续航功率下的耗油率为 238 g/(kW·h)。

4."隼"

1986 年建成的"隼"(СОКОЛ)级水翼护卫舰首舰,其靠左、右舷的两个动力装置是 M10Д 机组,中间的动力装置是 M16 机组。

M16 机组由 ДС71 燃气轮机和上、下两个直角传动齿轮箱组成,两个同轴定距桨安装在下部齿轮箱的轴上。ДС71 燃气轮机的功率为 7 355 kW,耗油率为 285 g/kW·h,倒车功率为 735 kW,翻修寿命为 10 000 h。

5."雕"

1988 年,大型反潜舰"雕"(Ястреб)级首舰交付苏联海军舰队。该舰的动力装置 M27

机组包括两台加速机组和两台巡航机组,其模式与 M15 机组的类似。巡航机组的 ДС71 燃气轮机功率为 6 620 kW,寿命为 30 000 h。

上述舰船的动力装置示意图如图 2 所示。

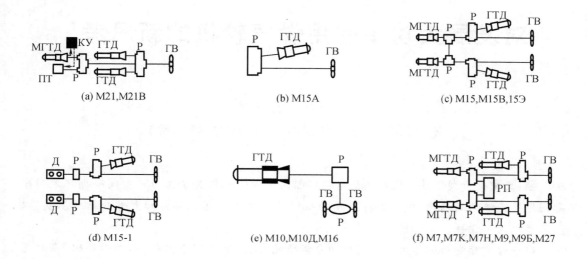

图 2　排水型水面舰船动力装置示意图

## 五、结束语

(1) 在近 30 年的发展历程中,针对 M70 系列燃气轮机在可靠性增长、气动性能改进、满足特定使用对象的需求、运行经验的积累和反馈、新技术应用等方面,相关研究人员做出了始终如一、持之以恒的努力,使之发展成一种成熟且具有广泛用途的机型。

(2) 通过灵活的动力组合方式,M70 系列燃气轮机的应用覆盖了气垫船、水翼艇、快艇、护卫舰、反潜舰、巡洋舰、天然气增压站和电站等广阔的领域。

(3) 燃气轮机的功率和使用寿命取决于所装舰船的类型和任务(载荷谱)。同一台燃气轮机应用于不同类型的舰船,其额定功率和寿命可能有较大差别,如一台在排水面舰船上使用寿命超过 10 000 h 的燃气轮机,若用于气垫船,其使用寿命可能仅为 1 000 h。

# 对发展大功率船用燃气轮机的新思考*

闻雪友　肖东明

(中国船舶重工集团公司第七〇三研究所)

**摘　要**：文中提出利用技术成熟的简单循环船用燃气轮机将其派生为一型间冷循环大功率船用燃气轮机的新构想，并且保持原发动机燃气发生器的通流部分和大部分结构不变，以继承原机的可靠性。

以一台实际发动机作为方案进行的分析研究表明，采用间冷循环，并精心匹配，在燃气轮机的功率显著增加(约28%)的同时，其效率也可略有提升(约3%)，具有工程实用价值。

## 一、大功率船用燃气轮机的发展趋势

由于燃气轮机属高技术产品，而且其研发必须具备雄厚的工业基础和长期投入，加之世界各国各大公司收购、兼并、联合，目前世界上真正能设计、制造船用大功率燃气轮机的厂商为数极少。各国海军装舰使用的燃气轮机集中在少数几个型号上。这一特点在大功率船用燃气轮机上尤为突出。进入20世纪90年代以来，大档功率(30 000马力[①]以上)燃气轮机的研发成为各国海军船用燃气轮机发展的支柱，见表1。

表1　大功率船用燃气轮机研发情况表(截至2006年)

| 型　号 | LM2500 | FT8 | UGT-25000 | WR-21 | LM2500+ | MT-30 | LM2500+G4 | LM6000PC |
|---|---|---|---|---|---|---|---|---|
| 样机投入运行的年份 | 1969 | 1990 | 1993 | 1997 | 1998 | 2001 | 2005 | 2006 |
| ISO条件下最大输出功率/hp | 33 600 | 36 860 | 42 400 | 33 850 | 40 500 | 48 275 | 47 370 | 57 330 |
| 热效率 | 0.372 | 0.389 | 0.381 | 0.421 | 0.391 | 0.398 | 0.393 | 0.421 |
| 耗油率/ [kg·(kW·h)$^{-1}$] | 0.227 | 0.217 | 0.221 | 0.200 | 0.215 | 0.212 | 0.214 | 0.200 |

\* 文章发表时间：2007年4月。

① 1马力(hp) = 0.745 699 9千瓦(kW)。

表1(续)

| 型号 | LM2500 | FT8 | UGT-25000 | WR-21 | LM2500+ | MT-30 | LM2500+G4 | LM6000PC |
|---|---|---|---|---|---|---|---|---|
| 压比 | 19.3 | 18.8 | 21 | 16.2 | 22.2 | 24 | 24 | 28.5 |
| 空气流量/(kg/s) | 70.4 | 83.3 | 87.6 | 73.1 | 85.8 | 116.7 | 92.9 | 123.9 |
| 制造公司 | GE | P&W | Zorya-Mashproekt | R.R | GE | R.R | GE | GE |

注：数据取自2007 GTW Handbook。

近15年来，大功率船用燃气轮机功率、效率的变化趋势见图1。从图1可知，大功率船用燃气轮机的最大功率有逐渐增长的趋势，其最大功率约为50 000马力(ISO)。同时效率也逐步提高，简单循环效率达到40%，复杂循环效率达到42%。

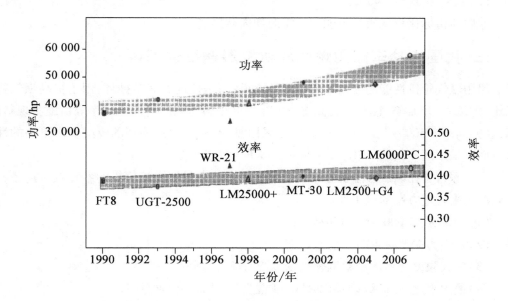

图1 大功率船用燃气轮机功率、效率的变化趋势

分析世界大功率船用燃气轮机的发展情况，可以发现在20世纪70年代以后相当长的时间段内，各国海军装舰使用的大功率船用燃气轮机绝大多数集中在LM 2500这一型号上。而进入20世纪90年代以后，随着各国海军对于水面舰艇主动力装置的功率需求不断提高，大功率船用燃气轮机的研发有了新的发展，陆续有FT8,UGT-25000,WR-21,LM2500+,MT-30等在研或已装舰，功率档主要集中在40 000马力左右。根据大功率船用燃气轮机的发展趋势可以推断，40 000～50 000马力档次的船用燃气轮机可以满足未来10～15年内各国海军对于大中型水面舰艇大功率燃气轮机主动力的需求。

## 二、我国大功率船用燃气轮机发展目标的建议

鉴于研制高性能船用燃气轮机的难度大、周期长、费用高的现实,在确定发展目标时应考虑如下因素:

(1)必须在一个合适的国产母型机基础上进行改型、发展或派生;
(2)应最大限度地保持高温部件不变,以继承其高可靠性;
(3)不追求全面的性能先进,但应体现总体性能的跃升;
(4)其性能指标应基本满足未来15~20年应用的需要;
(5)军民两用。

基于上述考虑,以及对国外大功率船用燃气轮机发展趋势的讨论,确定如下总体性能指标,在15 ℃,标准大气压力,不计进气损失、排气损失,无对外引气的条件下:

(1)功率为45 000~50 000马力;
(2)效率为38%~40%。

这两项指标比较现实且也能在一定时期内被接受。

## 三、我国大功率船用燃气轮机发展途径的探讨

船用燃气轮机性能的提高主要有两条途径:一是传统的简单循环,用于提高燃气初温、压比(加级)。增加空气流量,提高燃气发生器转子转速和改进部件效率等措施实现新的目标,这方面LM2500+是一个代表。二是采用复杂循环,通过循环的改进实现更高的性能,这方面以采用间冷回热(ICR)复杂循环的WR-21为代表。

文中探讨了在国产母型机基础上派生新的大功率船用燃气轮机的各种技术途径,并以下列准则来判定方案的适用性:

(1)功率为45 000~50 000马力;
(2)效率为38%~40%;
(3)最大限度地继承母型机燃气发生器的通流部分;
(4)最大限度地继承母型机部件的可靠性,尤其是高温部件;
(5)关键技术可在国内解决;
(6)较短的研制周期;
(7)较少的研制经费。

经过论证,文中提出采用间冷(IC)循环的新设想。

## 四、间冷循环燃气轮机

### 1. 间冷循环分析

间冷循环的原理图见图2。间冷循环燃气轮机的低压压气机(LPC)与高压压气机(HPC)之间设有一个中间冷却器(IC),使得空气在进入高压压气机前被冷却;空气经高、低

压压气机压缩进入燃烧室（CC），与燃料掺混燃烧后形成高温、高压的燃气，经高压涡轮（HPT）、低压 LPT 和与动力涡轮（PT）膨胀做功，分别驱动高、低压压气机和负载工作。与简单循环相比，间冷循环通过中间冷却器降低了高压压气机的入口温度，进而减少了其压缩空气的功耗，从而提高了发动机的输出功率。

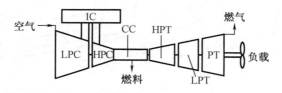

**图 2　间冷循环的原理图**

为了定量考察间冷循环燃气轮机的性能及主要循环参数间的关系，表 2 列出了计算分析所需的假定的各部件性能参数（均为现代大功率船用燃气轮机的较典型参数）。

**表 2　计算条件**

| 参　数 | 数　值 | 参　数 | 数　值 |
|---|---|---|---|
| $P_0/\text{bar}$① | 1.013 | $\eta_{CC}$ | 0.99 |
| $T_0/\text{℃}$ | 15 | $\eta_{HPT}^*$ | 0.87 |
| $T_{cool}/\text{℃}$ | 20 | $\eta_{LPT}^*$ | 0.89 |
| $\Delta P_{in}/\text{mmH}_2\text{O}$ | 100 | $\eta_{PT}$ | 0.92 |
| $\Delta P_{ex}/\text{mmH}_2\text{O}$ | 400 | $\xi_{IC}$ | 0.05 |
| $PR_{LPC}/PR_{HPC}$ | 1 | $\xi_{CC}$ | 0.05 |
| $\eta_{LPC}$ | 0.88 | $\xi_{ID}$ | 0.01 |
| $\eta_{HPC}$ | 0.90 | $LHV/(\text{kJ/kg})$ | 42 700 |

注：标 * 参数以 1 200 ℃ 为基准，初温每提高、降低 100 ℃，涡轮效率降低、提高 1.25%。

在不同的总压比（$PR = 12 \sim 42$）、不同的燃气初温（$TIT = 1\,100 \sim 1\,400$ ℃）和不同的间冷度（$\varepsilon = 0 \sim 0.85$）下，对发动机总体性能进行计算。

图 3 中对比简单循环和间冷循环的总性能图线，可以清晰地看出，间冷使燃气轮机的比功有大幅增加，无论是相应最大比功的压比值或是相应于最佳效率的压比值。在现代船用燃气轮机的参数下（燃气初温在 1 200 ~ 1 300 ℃，压比为 20 ~ 25），采用间冷对发动机耗油率的影响是负面的，间冷后，虽减少了压气机的功耗，增加了有效输出功率，但是压气机出口温度降低，为使燃气达到预定的初温值需要更多的燃料消耗。只有在很高的压比下，压气机功耗减少的影响会超过由于间冷导致燃料增加的影响，这时其耗油率会低于简单循环。

---

① 1 bar = 100 kPa（准确值）。

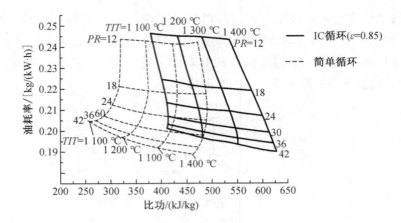

**图3 简单循环与间冷循环的性能曲线**

图4表示间冷对压气机出口温度的影响,这意味着进入高温涡轮叶片的冷却空气温度降低,在保持高温涡轮叶片金属表面温度不变时可允许适度提高燃气初温。此外,高压转子在压气机折合转速相同时,其物理转速也将明显减小,使其工作应力明显减小。

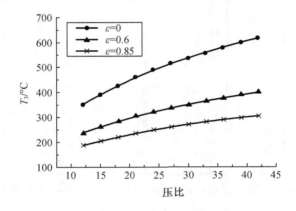

**图4 间冷对压气机出口温度的影响**

以上是间冷循环分析的简要结果。当把一台现有发动机作为发展为间冷循环燃气轮机的母型机而进行实际研究时,会发现一些有利于效率提高的因素,诸如燃气初温的适度提升、重新匹配后某些部件工作点效率的提高、折合转速的变化等,这些因素使得有可能在明显提高发动机功率的同时,其热效率也获得提升。

2. 间冷循环案例

选取一型发动机(MGT-33)作母型机用于发展为IC循环的研究,以讨论该方案的技术可行性。

MGT-33简单循环燃气轮机的主要性能见表3。

表3 MGT-33简单循环燃气轮机主要性能表

| 序号 | 参数 | 符号 | 数值(优化前) |
|---|---|---|---|
| 1 | 功率/kW | $N_e$ | 28 500 |
| 2 | 热效率/% | $\eta_e$ | 37 |
| 3 | 空气流量/(kg/s) | $G_a$ | 90 |
| 4 | 燃气初温/K | $TIT$ | 1 543 |
| 5 | 总压比 | $PR$ | 22.2 |

确定从MGT-33燃气轮机发展为IC循环燃气轮机(CGT-33)的主要原则如下:
(1) 在ISO条件下,CGT-33与母型机的燃气初温相同;
(2) 在ISO条件下,CGT-33低压压气机的折合转速与母型机相同;
(3) 最大限度地继承母型机燃气发生器的通流部分;
(4) 高压涡轮动叶不做改动;
(5) 最大限度地继承母型机部件的可靠性。

这些原则的贯彻实施,为改型后发动机的功率和热效率的提高提供了基本前提,并在母型机燃气发生器结构变动最小和通流部分通用性最高的前提下,继承了高可靠性。

经优化计算,获得循环性能见表4。MGT-33(上图)和CGT-33(下图)燃气轮机的示意图见图5。

表4 优化后CGT-33燃气轮机循环性能表

| 序号 | 参数 | 符号 | 数值(优化后) |
|---|---|---|---|
| 1 | 功率/kW | $N_e$ | 36 400 |
| 2 | 热效率/% | $\eta_e$ | 38.2 |
| 3 | 空气流量/(kg/s) | $G_a$ | 89.6 |
| 4 | 燃气初温/K | $TIT$ | 1 543 |
| 5 | 总压比 | $PR$ | 22.1 |

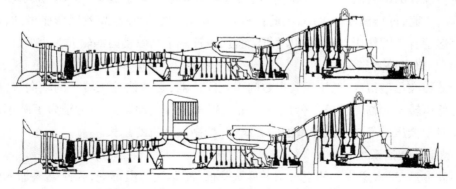

图5 MGT-33和CGT-33燃气轮机的示意图

该方案有以下 6 个技术要点：

(1) 压气机

低压压气机去掉末两级，并重新设计新末级导叶和校直叶片，使减级后的低压压气机的工作点在保持折合流量和折合转速不变的情况下，出口压力降低，平衡由于间冷造成的高压压气机入口折合流量降低的影响。调整后的低压压气机压比降低了 21%，效率提高了 2.41%。

高压压气机不变，匹配后的折合转速上升了 5.5%，物理转速降低了 13.2%，压比增高了 24.9%，效率降低了 1.7%。

(2) 中间冷却器

中间冷却器是大幅度提高功率的关键部件。由于在两个压气机之间增加了一个中间冷却器，使得空气在进入高压压气机前受到预冷而减少了高压压气机的功耗，有效地提高了发动机的功率。为实现高紧凑性、低流阻的中间冷却器，采用模块化结构，间冷度 0.85，间冷器的总压恢复系数为 0.95。

(3) 高、低压涡轮

鉴于母型机燃气初温较高，高压涡轮和低压涡轮叶片冷却结构复杂的特点，为了将技术风险降低到最小，高压涡轮动叶的结构保持不变。重新匹配后的高压涡轮虽因高压压气机功耗减小，使高压涡轮的膨胀比显著减小，但因圆周速度大幅下降，致使其载荷系数增加，导致涡轮效率下降 1.8%。

由于低压压气机拆除了末两级，因而低压涡轮膨胀比减小了 15.3%，效率提高了 0.91%。也因高压涡轮膨胀比减小，使进入低压涡轮的燃气温度增加了 3.3%。

由于进入高低压涡轮叶片的冷却空气温度显著降低，因而叶片金属表面温度仍不会超过母型机的相应值，加之高压转子的物理转速下降，导致离心力减小 24.7%。

(4) 动力涡轮

由于采用间冷及低压压气机拆除末两级，因此高、低压压气机的功耗减少，动力涡轮的入口温度增加 6.73%，膨胀比提高 32.1%，动力涡轮的输出功率显著增加，动力涡轮需重新设计。重新设计的动力涡轮效率提高了 1%，最终的功率增幅达到 27.8%。由于动力涡轮的效率远高于高、低压涡轮的效率，因此把相当一部分的能量移至高效区膨胀做功，对整机改善效率起了有效的作用，所以对 IC 循环而言，设计一个高效动力涡轮尤为重要。

(5) 低压转子临界转速

由于需要在母型机高、低压压气机之间引入间冷器，必然使低压转子的轴向尺寸增加，这将改变低压转子的临界转速。低压压气机去掉末两级的原因之一也是为了减小由于引入间冷器对于低压转子轴向尺寸增加的影响，并控制发动机的总长。

(6) 燃室与燃油系统

采用间冷后，进入燃烧室的空气温度显著降低，而燃气初温不变，因此燃料加入量增加（相对于简单循环增加了 24.1%），燃油泵、燃油喷嘴、燃烧室及燃油调节系统均需相应调整或重新设计。

## 五、结束语

（1）从一般的循环分析来看，采用间冷循环可使燃气轮机的比功明显增加，但对循环效率通常有负面影响。

（2）对实际发动机进行的 IC 循环改造方案分析表明，通过精心匹配，有可能在增大功率的同时提高效率。这是因为存在如下可供利用的因素：

①高压转子的物理转速明显下降，离心应力也显著减小，其折合转速有适度上调的可能；

②由于压气机出口空气温度明显降低，在保证高温涡轮叶片金属表面温度不高于规定值的前提下，燃气初温有上升的空间；

③由于燃气发生器的功耗减小，更多的能量转换在效率更高的动力涡轮部分完成；

④在燃气轮机重新匹配过程中，部分部件的效率、压比、膨胀比、温度等参数向有利方向变化。

案例分析结果表明，功率增加 27.8%，热效率提高 3.24%，发动机总体性能提高的效果显著。

（3）所研究案例的主要优点如下：

①母型机燃气发生器的核心部件结构变动最小，通流部分通用性最高，继承了高可靠性。

②与间冷回热循环相比，IC 循环规避了回热器及动力涡轮可调导叶两项关键技术，系统较简单，技术风险小，可明显缩短研发的周期，减少研发费用，又能使发动机总体性能上有所跃升。当然，也必须指出，其效率，尤其是部分负荷下的效率明显低于 ICR 循环。

③在功率和效率方面，可以满足未来 15~20 年内对大功率船用燃气轮机主动力的需求。

④间冷循环燃气轮机在国外舰船上也无应用先例，因此可以称之为"有中国特色的大功率船用燃气轮机"。

⑤所述研究将产生一个采用新循环的新机型，具有自主知识产权。

## 参 考 文 献

[1] WEN X Y, XIAO D M. Feasibility study of an intercooled cycle marine gas turbine[J]. Journol of Engineering for Gas Turbines and Power, 2008,130(2):022201.

## 附表  符号说明表

| 变量符号 | 变量说明 | 下标符号 | 符号说明 |
| --- | --- | --- | --- |
| $P$ | 压力 | in | 入口 |
| $T$ | 温度 | LPC | 低压压气机 |
| $\Delta P$ | 压力损失 | IC | 间冷器 |
| $PR_{LPC}$ | 低压压气机压比 | HPC | 高压压气机 |
| $PR_{HPC}$ | 高压压气机压比 | CC | 燃烧室 |
| $PR$ | 总压比 | HPT | 高压涡轮 |
| $TIT$ | 燃气初温 | LPT | 低压涡轮 |
| $Ne$ | 功率 | PT | 动力涡轮 |
| $SFC$ | 油耗率 | ID | 中间扩压器 |
| $Ga$ | 流量 | ex | 排气管 |
| $\eta$ | 效率 | cool | 冷却剂 |
| $\xi$ | 压力损失系数,$\xi = \Delta P$ | | |
| $\varepsilon$ | 间冷度,$\varepsilon = (T_2 - T_{22})/(T_2 - T_{cool})$ | | |
| $\eta_e$ | 循环效率 | | |
| $SP$ | 比功 | | |
| $LHV$ | 燃料热值 | | |

# 现代舰船燃气轮机发展趋势分析*

闻雪友　肖东明

(中国船舶重工集团公司第七〇三研究所)

**摘　要**：通过对近20年世界船用燃气轮机发展趋势的分析，指出目前船用燃气轮机的发展重点是大功率船用燃气轮机，机型也更集中，其次是小档功率船用燃气轮机。功率 29.4～36.75 MW、效率39%～42%的船用燃气轮机是未来10～15年各国海军采用的主要机型。新研机型与综合电力推进系统相结合的特点非常明显。

## 一、引言

目前，船用燃气轮机已广泛应用于各类水面舰艇及民用船舶，船用燃气轮机功率密度高、机动性好等一系列优点在实际应用中已形成共识。文中对近20年(1991—2010年)世界舰船燃气轮机的发展状况进行分析，旨在对我国船用燃气轮机的发展有所借鉴。

## 二、船用燃气轮机概况

（一）船用燃气轮机的应用概况

图1给出了船用燃气轮机的应用分类情况。按装船的燃气轮机台数计，前四位依次为驱逐舰、护卫舰、气垫船、巡洋舰；按装船的燃气轮机总功率计，依次为护卫舰、驱逐舰、气垫船、快艇；按装燃气轮机的船数计，依次为护卫舰、驱逐舰、巡洋舰、快艇。从统计结果可见，驱护舰、气垫船、高速艇是船用燃气轮机的主要市场。船用燃气轮机在军船上的应用数量明显大于民船应用数量。

---

\* 文章发表时间：2010年8月。

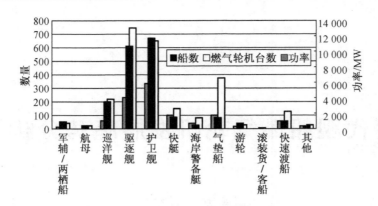

**图 1　船用燃气轮机推进市场的应用情况**

(二) 新研船用燃气轮机分析

图 2 给出近 20 年(1991—2010 年)来新研的船用燃气轮机的情况,图中纵向分为小档(1.47～11.025 MW)、中档(12.495～22.05 MW)和大档(23.52 MW 以上)功率船用燃气轮机,横向分成了前 10 年(1991—2000 年)和后 10 年(2001—2010 年)。从图中可得出以下结论:

(1) 近 20 年来,新研大档功率船用燃气轮机 6 型,中档功率船用燃气轮机 3 型,小档功率船用燃气轮机 9 型。可明显看出,世界范围内船用燃气轮机的发展重点是"一大"(大功率船用燃气轮机)和"一小"(小功率船用燃气轮机),尤其是大功率船用燃气轮机,也反映出实际的市场需求和技术需求。

(2) 对比前 10 年和后 10 年,新研大功率船用燃气轮机由 4 型降为 2 型,中档功率船用燃气轮机由 3 型降为 0,小档功率船用燃气轮机由 7 型降为 2 型,也印证了前述的发展重点是"一大一小"的观点,同时也反映出世界范围内所使用的燃气轮机的机型更集中的特点,更便于运行、维护、保养和降低全寿命成本。

(3) 上述特征在一定程度上也与现代舰船采用柴油机电力和燃气轮机联合动力装置(CODEAG),柴油机电力和燃气轮机交替动力装置(CODEOG),燃气轮机电力和燃气轮机交替动力装置(COGEOG),燃气轮机电力和燃气轮机联合动力装置(COGEAG)和燃气轮机和综合电力推进系统(IEP)的趋势有关。

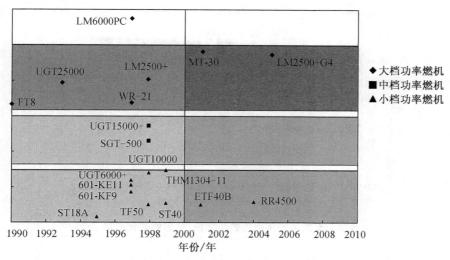

图 2　1991—2010 年新研船用燃气轮机一览图

（三）大功率船用燃气轮机性能发展趋势

表 1 列出了现代大功率船用燃气轮机的主要性能（文中所表述的船用燃气轮机性能均是在大气温度 15 ℃，不计进排气损失条件下的值）；图 3 给出了功率、效率的变化趋势。很明显，单机功率有逐渐增大的趋势，目前，最大功率为 352.8 MW。热效率也正稳步提高，简单循环船用燃气轮机已达 40%，复杂循环船用燃气轮机已达 42%，相应地最高压比达 24；高压涡轮动叶入口最高温度达 1 240 ℃。需指出，表 1 中所列 LM6000PC 的功率、效率均很高，但据了解迄今尚无军方用户订货。可以预期，功率 29.4～36.75 MW、效率 39%～42% 的大功率船用燃气轮机将是未来 10～15 年内海军舰艇采用的主力机型。

表 1　大功率船用燃气轮机研发情况表（1991—2010 年）

| 型　号 | UGT-25000 | LM6000PC | WR-21 | LM2500+ | MT-30 | LM2500+G4 |
|---|---|---|---|---|---|---|
| 样机投入运行的年份/年 | 1993 | 1997 | 1997 | 1998 | 2001 | 2005 |
| ISO 条件下最大出力/MW | 28.89 | 42.14 | 24.88 | 29.77 | 35.48 | 34.82 |
| 热效率 | 0.374 | 0.421 | 0.421 | 0.391 | 0.398 | 0.393 |
| 耗油率/(kg/(kW·h)) | 0.221 | 0.200 | 0.200 | 0.215 | 0.212 | 0.214 |
| 压比 | 21 | 28.5 | 16.2 | 22.2 | 24 | 24 |
| 空气流量/(kg/s) | 87.6 | 123.9 | 73.1 | 85.8 | 116.7 | 92.9 |
| 制造公司 | Zorya-Mashproket | GE | RR | GE | RR | GE |

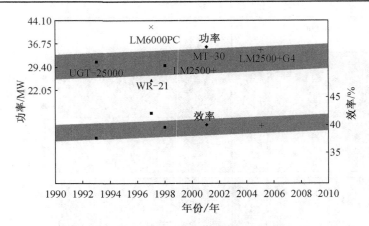

图 3 大功率船用燃气轮机功率、效率的变化趋势

## 三、新机型技术特点分析

（一）LM2500 + /LM2500 + G4

LM2500 船用燃气轮机是美国 GE 公司发展最为成功的一型船用燃气轮机,在 1969 年问世时,功率为 16.54 MW,效率为 36%,经 40 年的发展成为功率 34.82 MW,效率达 39.3% 的新机型,形成 LM2500/LM2500 + /LM2500 + G4 的系列(图 4)。LM2500 成为国际舰船燃气轮机市场中销售最持续旺盛的发动机。

LM2500 船用燃气轮机从航空发动机 TF39/CF6 派生而来(图 5),保留了原航空发动机的核心机,去除了低压风扇,将驱动低压风扇的原低压涡轮改为动力涡轮,这种改动方式最大限度地继承了该航机系列已积累的上千万运行小时的可靠性,使 LM2500 船用燃气轮机在较短时间内就建立起良好的信誉。其后又在压气机前加零级,高压涡轮采用新材料和新的冷却结构,重新设计动力涡轮的叶型,使其发展为性能更好的 LM2500 + (图 6)。

此后,在 LM2500 + 的基础上又通过对高压压气机、高压涡轮和动力涡轮的部分叶型及材料做了调整,增加了通流能力,能承受更高的温度,使总体性能又上了一个新台阶。

LM2500 系列的发展模式堪称经典。

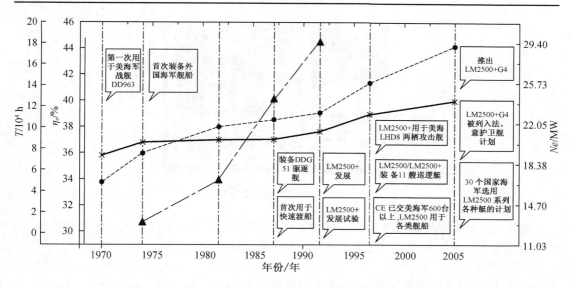

图 4　LM2500 船用燃气轮机的发展历程

图 5　LM2500 燃气轮机纵剖图

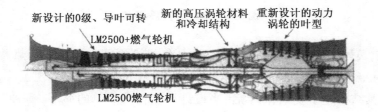

图 6　LM2500 + 燃气轮机纵剖图

（二）WR – 21/MT30

这是英国罗尔斯 – 罗依斯公司（R.R）新研的 2 型各有特色的船用燃气轮机。

WR – 21 间冷回热循环燃气轮机（图 7）的特点是：采用间冷提高功率，采用回热和动力涡轮首级可调导叶提高效率，不仅额定工况具有高效率（42%），而且在 1/3 额定负荷下仍有接近额定值的效率，特别适合舰船的要求。该机现装于英国 45 型驱逐舰和"伊丽莎白女皇"新航母的综合电力推进系统中。

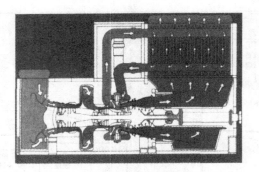

图 7　WR-21 燃气轮机装置纵剖图

WR-21 在研制过程中曾得到众多国家海军的追捧,各国海军纷纷表示了采购意向,但值得注意的是,当 WR-21 正式面世时并没有获得预期的热烈反响。

与采用复杂循环的 WR-21 不同,MT30 是简单循环(图 8)。MT30 的基本设计由 Trent 800 航空发动机缩小而来,有 1 个 8 级可变几何的低压压气机,高压压气机为 6 级,动力涡轮 4 级,功率 35.48 MW,效率 40%。MT30 已装备于美国海军的濒海战斗舰,也以发电模块方式与 WR-21 共同构成英国新航母的综合电力推进系统。

图 8　MT30 船用燃气轮机

50 多年前,英国 MV 公司设计了包括 3 个压气机、2 个间冷器、1 个回热器、1 个燃烧室、3 个涡轮的船用燃气轮机 RM60。该燃气轮机装于"灰鹅"号炮艇,实现了改善低工况性能的目标。由于其结构过于复杂,4 年后,当设计性能相当简单的循环"海神"发动机在"勇敢"级快艇上成功运行后,RM60 便不再受到青睐。

事物是螺旋上升的,现在 R.R 公司的型号为 WR-21 和 MT30 的 2 型发动机的同时应用又为研究人员提供了一个极好的观察机会。

(三)LM6000PC

前已提及,LM6000PC 目前并未在军船上应用,但在民船,例如近海钻井泊船上有应用。该机的示意图见图 9,属于双转子套轴燃气轮机,其独特之处在于低压转子直接驱动,没有动力涡轮。低压压气机 5 级,压比 2.4;高压压气机 14 级,压比 12.1,总压比 30。高压涡轮 2 级驱动高压转子,低压涡轮 5 级驱动低压压气机和负荷,负荷可加在低压转子的冷端或热端。在这样的负荷下为扩大稳定运行范围,在压气机上采用了3项可变几何措施:进口可调导叶、高低压压气机间有可调旁通阀、高压压气机前5级可调。每个系统单独控制,提高了发动机控制的灵活性,使其能在运行范围内免于失速。应该指出,由于低压涡轮与低压压气机和负荷直接相关联,因此,运行中当低压转子转速为0时,该燃气轮机扭矩将会急升。

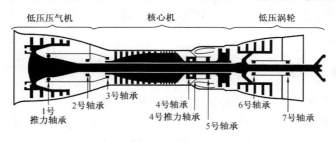

图 9　LM6000PC 燃气轮机示意图

### (四) RR4500

4.41 MW 的 RR4500 是 R.R 公司专为美国海军 DDX 计划综合电力系统中的船用发电机组开发的,为此,其与 MT30 燃气轮机发电模块共同在美国费城海军试验基地进行了综合电力推进的运行试验。

RR4500 基本上是 3.89 MW Allison 501(用于 DDG51 的船用发电装置)的技术升级版,其属于 1 个单轴、高压比的燃气轮机,功率为 4.43 MW,效率为 32.25%。

## 四、适应综合电力推进的燃气轮机试验装置

为了研发和进行耐久试验,使燃气轮机试验装置既可用于发电运行模式又可用于机械驱动运行模式,国外新建了试验站(图10)。试验站有三个特点:一是可以试验燃气轮机发电机组,包括发电机及控制系统,也可模拟机械驱动负荷(3次方规律)运行;二是用一系列的电阻器负载组来消耗发电机所产生的能量;三是利用电阻器负载组产生的热空气导入燃气轮机进气装置,使发动机能在接近规定的进口空气温度下运行,这点对验证试验尤为重要。

图 10　MT30 燃气轮机试验站

鉴于美国船级社(ABS)规定:耐久试验在38℃下进行,试验中如果不采用进口空气加热试验装置,在环境温度低于38℃时就需提高功率设定值,以产生当量(相当于38℃下)的涡轮热部件温度。

MT30燃气轮机试验站在耐久性试验装置的控制系统中设置了1个自动负荷控制台,可以自动执行试验。

## 五、结束语

文中对现代船用燃气轮机发展趋势进行了分析,可见:

(1)目前,船用燃气轮机的发展重点是"一大"(大档功率船用燃气轮机),"一小"(小档功率船用燃气轮机),尤其是大功率燃气轮机,机型更集中。

(2)功率29.4~36.75 MW,效率39%~42%的大档功率船用燃气轮机将是未来10~15年内海军舰艇采用的主力机型。

(3)经过40年来持续不断的改进、发展,LM2500系列燃气轮机无论在技术发展还是市场应用方面均堪称经典。

(4)尽管新研机型(WR-21,MT30,LM2500+,RR4500)各具特色,但均已与综合电力推进系统结下不解之缘,相应地出现了可用于发电模式及机械驱动模式的新试验装置。

## 参 考 文 献

[1] FARMER R F. LM2500 upgrade nominally rated at 46 000 shp and 41% efficiency[J]. Gas Turbine World,2005,(6):10-14.

[2] 闻雪友,李伟. WR-21:新一代的船用燃气轮机[J]. 热能动力工程,1999,14(1):1-6.

[3] BRANCH D, WAINWRIGHT J. Development and qualification of the marine trent MT30 for next generation naval platforms[C]//ASME Turbo Expo 2007: Power for Land, Sea, and Air. American Society of Mechanical Engineers Digital Collection, 2007: 947-951.

[4] SIDENSTICK D, MCANDREWS G, TANWAR R, et al. Development, testing, and qualification of the marine LM6000 gas turbine[C]//Turbo Expo: Power for Land, Sea, and Air. 2006, 42401: 47-53.

[5] 闻雪友,肖东明. 对发展大功率船用燃气轮机的新思考[J]. 舰船科学技术,2007,29(4):17-21.

# 船用燃气轮机 GT1000 *

闻雪友　赵友生

(哈尔滨船舶锅炉涡轮研究所)

**摘　要**：GT1000 船用燃气轮机由航空 SPEY MK202 双转子内外涵混合加力涡扇发动机派生而来。文中详细描述了 SPE YMK202 发动机的船用化改装和 GT1000 机的主要技术特点，包括气动、结构设计、零部件强度和系统布置等，也给出了 GT1000 整机的性能试验结果。

## 一、引言

20 世纪 70 年代，我国从英国罗尔斯－罗伊斯公司引进 SPEY MK202 双转子涡轮风扇内外涵混合加力式航空发动机，供航空部门使用。

航空斯贝发动机有民用和军用两个系列，经各种机型的发展与改进，其性能有了很大的提高。燃气初温由 1 313 K 提高到 1 440 K，总压比由 15.4 增高到 21.4，内涵流量由 46 kg/s 加大到 56～68 kg/s。到 1981 年底，该发动机使用总数已达到 4 400 台，累计运行 2 200 万小时，发展得相当成熟，因而其是一型适用于船用化改装的功率等级适中的机型。英国罗尔斯－罗伊斯公司与美国库帕－罗尔斯工业公司已先后将航空斯贝 RB168－66 型 (TF41－AQ) 发展成 5 型工业及船用燃气轮机：库帕－罗尔斯工业斯贝燃气轮机、SMIA 船用燃气轮机、SK15HE 燃气轮机热电联供装置、SMIC 燃气轮机、SMCICR 燃气轮机[1]。

航空改装技术是一种可用最快和最节省的方式，使发动机在工业应用中保持领先地位的航空发动机技术。高性能发动机的设计、研究、运行验证是一个长期且花费巨大的过程，航空改装在财政上的优势更使其受到强烈偏爱。为扩大应用范围，实现一机多用，研究人员开始了一项将国产斯贝 MK202 发动机改装为船用燃气轮机的研究计划。

## 二、船用化途径

英国在 10 多种斯贝型号中选择了 RB168－66 型 (TF41－A－2) 作为工业/船用改装的母型机，分析其原因主要有如下三点：

---

* 文章发表时间：1997 年 1 月。

(1) 其核心机流量大,因而改装后功率大。

(2) 该机型原为美国海军 ATE"海盗"Ⅱ歼击机设计,有关部件已考虑了抗腐蚀要求,因此特别适用于船用改装。

(3) 风扇级数少(两级)对实施顶切方案(工业型)有利。

我国引进的是 MK202,其内涵流量比 RB168-66 小 17.6%,最大推力也明显小于前者。MK202 有五级风扇,这对将风扇顶切成低压压气机的方案来说,难度增加。

基于上述情况,MK202 的船用化改装有两个基本方案:

(1) 燃气发生器将 MK202 发动机的原低压风扇在适当的半径部件上做顶切,使之成为低压压气机,原低压涡轮做相应的匹配调整。

(2) 燃气发生器完全舍弃低压风扇,重新设计一个低压压气机,对低压涡轮也需做相应的匹配调整。

对每一个燃气发生器相应地设计一个动力涡轮机,构成船用燃气轮机。

第一种方案(A 型)的优点是低压压气机的结构继承性强,改装工作量相对较小;不足之处是功率相对较小,耗油率较高,变工况性能较差,五级风扇顶切难度较大。

第二种方案(B 型)的优点是功率较大(重新设计的低压压气机为 6 级,压比升高,空气流量增大),耗油率较低,变工况性能较好;不足之处是重新设计低压压气机并使之与原高压压气机协调工作的难度较大,新结构的可靠性亦需验证。

根据当时实际使用对象的需要,具体设计时分两步走,先实施 A 型,再过渡到 B 型。为此,在动力涡轮的设计上需充分考虑两种燃气发生器的通用性问题。

文中主要介绍 A 型发动机(GT1000)的情况。图 1 的上半部分为发动机纵剖面图,下半部分为原航空发动机低压风扇和外涵道图。

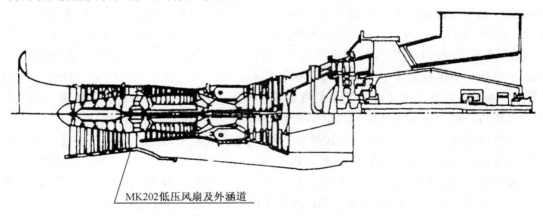

MK202低压风扇及外涵道

图 1　GT1000 燃气轮机纵剖面图

## 三、技术要点

斯贝 MK202 发动机船用化改装的设计点是在寿命、效率、变工况性能间进行了全面的权衡后确定的。最大连续工况的燃气初温比原航机的燃气初温降低了 115 ℃,转速、压比、流量也均相应降低。

对低压风扇的顶切部位,应在设计流量下原机内外涵道的分流界面,基本上是按通流面积缩小约40%的所在半径,沿三元流计算的流线方向,并适当考虑实际修正而切去风扇的外涵道,构成一个5级轴流式低压压气机。为保证中、低转速范围内工作可靠,进口导叶可调。低压压气机的试验特性图见图2。重新设计的低压压气机(用于B型)的试验特性图完全证实了压比提高、变工况性能改善的预期目标,如图3所示。

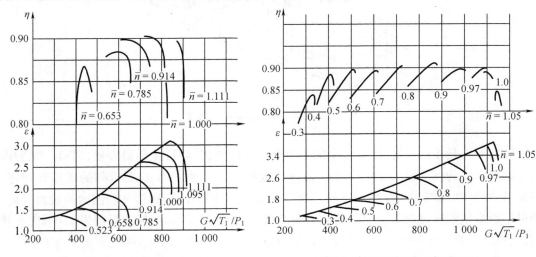

图 2　A 型低压压气机特性图　　　图 3　B 型低压压气机特性图

风扇顶切成低压压气机功耗减少,因此设计时对原低压涡轮的导叶和动叶进行了更改,使低压部分协调工作。

12级轴流式高压压气机和带动其两级轴流式高压涡轮组成高压转子,为保持压气机在低转速下稳定工作,高压压气机采用了进口导叶可调和7级放气机构,高、低压两转子为套轴结构。

燃气发生器所有叶片均涂有耐蚀涂层。

考虑海洋工作条件和长寿命的要求,10个火焰筒和燃气导管内、外表面用等离子喷涂镍钴铬钇防高温硫化腐蚀涂层,其内表面还复合涂锆碳镁涂层。

燃气发生器通过前后两个支架安装在基座上。发生器的后支架可承受轴向力,为发生器的死点,其允许发生器的安装边与后支架有径向的相对膨胀。发生器的前支架是一个可摆动的、单侧的支板,能满足发生器向前自由膨胀的要求,并能减弱水下爆炸冲击对发动机的影响(图4)。

燃气发生器的前端与进气弯管相连接,后端经中间扩压器与动力涡轮做软连接,可减少两端振动的交互影响。燃气发生器与动力涡轮各自分别借由两个机架安装在一个钢制整体基座上,因此燃气发生器可以单独快速换装。

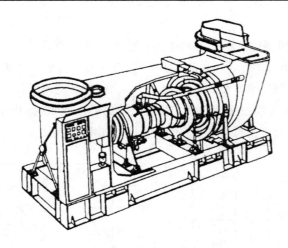

图 4　GT1000 燃气轮机箱装体图

动力涡轮是按照长寿命、高效率的思想进行设计的，气动负荷选得较低，留有一定的潜力。动力涡轮的气动设计及结构设计还仔细地考虑了 A 型机和 B 型机间的通用性问题。在更换叶片（扭曲规律及叶型均相同，仅叶高不同）、中间扩压器和排气管前面的过渡段后，即可实现型号间的过渡。动力涡轮的功率极限为 110％ 额定功率，转速极限为 110％ 工作转速，扭矩极限为 145％ 额定扭矩。冲击值是按垂向 $40\ g_n$、横向 $25\ g_n$、纵向 $6\ g_n$ 考虑的。

两级轴流式涡轮转子采用悬臂式安装在轴承座上，使得两个轴承均工作在温度较低的涡轮排气侧，且易于靠近、检修。轴承采用可倾瓦径向轴承和直接润滑的滑动推力轴承。轴承座和轴承设计成可通用于左旋或右旋的涡轮转子，滑油的进、排油管的布置能通用左舷或右舷布置的需要调整。

动力涡轮采用整环结构，以减少受热后的膨胀不均匀性。转子在船上即可用专用工具方便地拆除，便于检修，更换叶片等零件。导向器亦可在船上方便地吊出。

为满足动力涡轮运行转速范围宽广的要求，转子设计成刚性转子。为了改善气动和振动性能，动叶采用带 Z 形冠叶片。

动力涡轮后支撑的"内"传力结构，即使在受到较大的冲击载荷的情况下也易于保持动、静部件间的间隙（相对于悬臂式的"外"传力结构）。

整个动力涡轮通过两个支架单独地安装在基座上。前支架为死点，仅允许径向热胀，后支架允许动力涡轮向后膨胀。

排气管的前端刚性地安装在动力涡轮上，排气管的后端则有其单独的支架，并可自由地向后膨胀。

箱装体提供了隔音、隔热和三防保护，利用排气引射进行冷却，还提供了机旁控制装置和部分辅助设备，并对燃油、滑油、空气、水、供电和控制电缆提供了接口。

燃油系统的主要变动是采用轻柴油作燃料。

滑油系统包括燃气发生器及动力涡轮两个系统，两者均为反向循环系统，滑油冷却器装在供油路上。

调控系统是在原 MK202 发动机主燃油控制系统的基础上充实、发展而成的,主要是增加了动力涡轮极限转速控制、超速保护装置和燃料开关(实现超速、振动超值、火警及滑油压力过低的保护功能)。

## 四、性能

虽然 GT1000 燃气轮机的燃气发生器是由斯贝 MK202 发动机改进而成的,但实际上,发生器零组件总数的 30% 是新设计件[2]。研制中,仅燃气发生器部分所进行的零部件试验、附件试验、全台燃气发生器试验就超过 70 项。动力涡轮及箱装体亦是新研制的。因此,在进入整机试验前进行了大量的有关空气动力学、燃烧、传热、强度、振动、结构应力分析、声学特性、附件、材料、防腐、热冲击、涂层、环境条件等试验,并在此基础上进行了整机性能试验。

在大气压为 0.101 3 MPa,大气温度为 15 ℃,不计进、排气损失的条件下,GT1000 燃气轮机的性能如表 1 所示。

表 1　GT1000 燃气轮机性能表

| 性　能 | 数　值 |
| --- | --- |
| 最大持续功率/kW | 10 500 |
| 耗油率/(g/(kW·h)) | 265(燃油低热值 42 705kJ/kg) |
| 空气进气流量/(kg/s) | 约 50 |
| 压比 | 约 15.5 |
| 燃气初温/℃ | 1 052 |
| 箱装体尺寸 | 7 800 mm × 2 500 mm × 3 200 mm |
| 总质量/t | 25 |

发动机功率随大气温度变化的关系见图 5,发动机耗油率与功率的关系曲线见图 6。

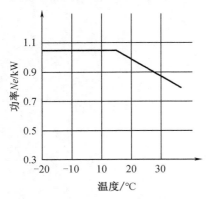

图 5　GT1000 功率随大气温度的变化

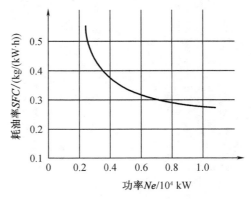

图 6　GT1000 的性能曲线

## 五、结束语

由航空斯贝 MK202 涡轮风扇发动机派生而成的 GT1000 船用燃气轮机,总体设计合理,机组运行平稳,保护系统可靠,主要技术指标达到了预期的目标。这也表明国产航空斯贝 MK202 发动机是一台实用的母型机,其工业/船用系列化发展是符合国情的,技术上是可行和现实的,有一定的先进性。

GT1000 的研制工作由哈尔滨船舶锅炉涡轮机研究所和西安航空发动机公司联合实施。

## 参 考 文 献

[1] 闻雪友,赵友生.国产航空斯贝发动机的工业及船用化发展[J].热能动力工程,1988,3(4):1-8.
[2] 金理斌.论国产斯贝系列航机改型燃气轮机的开发[J].燃气轮机技术,1993,6(4):16-22.

# WR-21——新一代的船用燃气轮机*

闻雪友　（中国船舶重工集团公司第七〇三研究所）
李　伟　（海军装备部）

**摘　要**：WR-21是21世纪初的新一代船用燃气轮机。其采用了间冷、回热技术，具有优良的变工况性能。文中综述了该机的研制历程，发动机的性能，各主要部件的设计和维护特点。

## 一、引言

近年来，船舶及工业燃气轮机领域的人们总是情不自禁地把关注的目光投向WR-21燃气轮机的研制工作。WR-21是一种带有中间冷却器和回热器的燃气轮机，与简单循环相比（如LM2500），不仅输出功率增加，额定功率下的耗油率降低，而且在大部分功率范围内具有平坦的耗油率曲线，同时又具有低噪声及低的排气红外特征。与美国海军现役的燃气轮机推进动力水面舰船相比，采用该动力装置的舰船年燃油消耗量降低约30%，成为新一代船用燃气轮机的象征。

## 二、历程

约50年前，船用燃气轮机的先驱——英国，认识到船用燃气轮机要求在整个功率范围内具有低的耗油率（尤其是部分负荷）。为实现这一点，相关人员尝试选择了包括两个压气机、两个中间冷却器、一个回热器、三个涡轮的复杂循环，即MR60燃气轮机。其于1951年进行了陆上试验，功率达到3 898 kW（5 300马力）。作为船用燃气轮机，这个装置还是值得注意的，因为其尽管装置复杂，仍然达到了3.59 kg/kW（2.64 kg/hp）的比重，其在40%~80%功率范围内的耗油率曲线平坦。1954年，该燃气轮机安装于"灰鹅"号炮艇并进行了海上航行，证实了其具有可靠性。然而RM60虽然达到了良好的部分负荷性能，却使发动机

---

\* 文章发表时间：1999年1月。

的购置成本大大提高,且因其体积庞大,控制复杂循环系统的损失也是困难的,再加上其性能很快即被改进的简单循环发动机("海神")所超越,因而 M60 并未获得进一步发展。

几十年后,由于燃气轮机及热交换器方面的技术进展,使间冷－回热循环在达到最高的循环效率、优良的变工况性能的同时仍能使装置结构紧凑,这使其可能特别适于舰用。

1981 年,Rolls－Royce 公司向美国海军提出发展间冷－回热燃气轮机(ICR)的设想。

1985 年秋,美国海军首次招标 38 ℃,17 897 kW ICR 高效舰用燃气轮机。Rolls－Royce 公司与 Allison,Garrett 公司合作提出了 TF41/Sper 改型方案,并与 GE M&I 提出的 LM1600 改型和 P&W 的 PW2037 改型方案先后于 1985 年底和 1986 年得到一年的概念设计合同。1986 年续一年预研合同。

1987 年底,美海军决定将功率提高到 19 686 kW 并考虑全电力推进方案,至 1989 年底,美海军也只是维持 GE 及 Rolls－Royce 公司的预研工作。

1990 年,美海军恢复 ICR 计划,重新招标,确认 19 686 kW 并有 10% 的增长潜力,尺寸适合于 LM2500 换装。Rolls－Royce 公司与 Westinghouse 公司合作,结合 RB211 系列航空发动机单元体结构,获得了与航空型零件高度通用的最佳 ICR 方案。

1991 年 12 月,美国海军将 ICR 燃气轮机机组的设计和发展合同授予 Westinghouse Electric Coporation(WEC)船舶分部,分承包商主要有 Rollse－Royce 公司工业与船用燃气轮机分部(负责燃气轮机),Allied Signal 公司航空系统和设备集团(负责回热器和中间冷却器)和 CAE 电子公司(负责控制设备)。

1991—1995 年为前期发展阶段(设计、发展、调试和确认主要性能参数)。

1994—1999 年为全尺寸发展阶段(美国海军试验、调试、训练、编写说明等)。

1997 年 5 ~ 9 月,由 Northrop Grumman 船用系统公司(新的美国总承包商)与 RR 公司共同在英国国防评估和研究机构的 Pyestock 试验场完成了产品型标准系统的最新 500 h 耐久性试验,共试验 503 h,69 次启动,相当于在海上实际运行 1 500 ~ 2 000 h。预计 1998 年美国海军在 Philadelphia 的海军水平舰船中心的 Carderock 分部建的第二个试验场上进行第二台机的试验。

原计划在 1999 财政年度分批装备 DDG－51"Arliegh Burke"级导弹驱逐舰。但按美国海军目前计划,第一台间冷－回热燃气轮机要到 2004 财政年度才装备于战舰,比预计的晚 5 年,装备最后 9 艘 DDG－51 级舰。

1994 年 7 月,首台前期发展型 WR－21 开始试验以来,多台 WR－21 试验型、发展型以及生产型样机已相继完成各项性能试验和耐久性试验。至 1997 年,总试验时数已超过 1 200 h,其中包括 110% 负荷的重要试验。在首台样机试验前完成了各模块和部件试验。预计 3 000 h 验收试验将在 1999 年进行。WR－21 表现出的高性能已引起世界各国海军的极大兴趣,目前考虑采用此新型动力装置的舰船有:美国海军 DD－21 级驱逐船、DDG－51 级驱逐舰,英国新型航母(以替代 2010 年退役的 3 艘"无知"级轻型航母),英国、法国、意大利"地平线计划"中的新一代"前景"级护卫舰(CNGF)。为此,英国和法国海军根据 1994 年 6 月和 1995 年 9 月分别签订的协议备忘录,负担该计划项目的部分经费。

## 三、设计要求

简单循环与 ICR 循环的线图见图1。

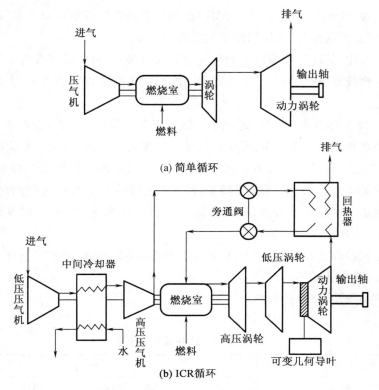

图1　简单循环与 ICR 循环线图

WR-21 的纵剖面图见图2。

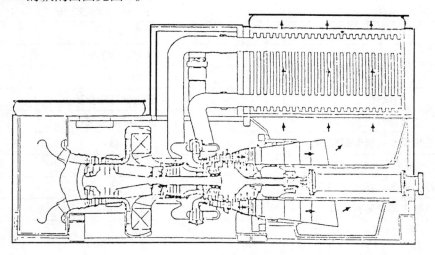

图2　WR-21 的纵剖面图

ICR 循环与简单循环燃气轮机的不同之处在于：

（1）ICR 循环在两台压气机之间设有一台中间冷却器，使得空气在进入高压压气机前被冷却，从而减少高压压气机所需的功耗，改进高压轴的效率并增加发动机的功率（在 WR-21 上约增加 25% 的功率）。此外，间冷也降低了高压压气机出口的空气温度，提高了空气与燃气排气温度的温差，从而增加了回热器的效率。

（2）ICR 循环中，高压压气机出口的空气先进入回热器，利用燃气排气余热对空气进行预热，然后进入燃烧室，这就降低了为达到预定的燃气初温所需的燃料量，实现了在耗油率方面的重要改进。

（3）从燃气发生器出来的燃气，通过第一级可变几何导叶（VAN）进入动力涡轮。随着功率减少，动力涡轮质量流量减少，可变几何导叶逐渐关小。因此，对一给定的部分负荷可保持高的燃气初温（在 WR-21 中，全工况和 30% 工况下动力涡轮入口燃气温度均为 852 ℃；出口温度则从全工况的 355 ℃ 降为 30% 工况时的 272 ℃），这使回热器空气/燃气的温差增加，改进了回热器的传热，使部分负荷下的耗油率获得改善。在 WR-21 上回热加 VAN 可能节省 30%～40% 的燃油。

回热器、中间冷却器和涡轮可变几何导叶是使 WR-21 相对于简单循环获得耗油率减小和功率增大的三个关键特征。

WR-21 的主要技术要求见表 1。

**表 1　WR-21 主要技术要求**

| 项　目 | 技　术　要　求 |
|---|---|
| 功率/MW | 19.4(26 400 马力) +10% |
| 耗油率/(g/(kW·h)) | 222.4(163.6 克/(马力·小时))(100% 功率)<br>233.6(171.8 克/(马力·小时))(30% 功率) |
| 尺寸 | 8 m × 2.64 m × 4.83 m |
| 质量/t | 54.55 |
| 可靠性/h | 1 000(MTBF) |
| 寿命/a | 40 |
| 模块化 | 主要组件可单独更换 |
| 主要部件更换 | 通过船舶进气道 |
| 燃气发生器平均换装时间/h | 48 |
| 动力涡轮平均换装时间/h | 72 |
| 中间冷却器平均换装时间/h | 24 |
| 回热器平均换装时间/h | 120 |

表1(续)

| 项　目 | 技术要求 |
|---|---|
| 控制 MTTR/h | 2 |
| 润滑油单元 MTTR/h | 9.75 |
| 可维修 | 适于船上使用 |
| 船上维护(O 和 I 级水平)MTTR/h | 24 |
| 船员预防性维护(计划)/(h/周) | 4.5 |
| 舰员维护(计划和非计划的)/(h/周) | 6.75 |

注：MTTR 指平均修复时间，即 Mean time to repair；

MTBF 指平均故障间隔时间，即 Mean time between failure。

表1中的功率和耗油率是在美国海军规定的条件下的性能值：大气温度38 ℃，海平面，湿度40%，进气损失0.98 kPa，排气损失1.47 kPa，海水温度29 ℃。

WR-21在ISO条件下全工况性能(对于发展成熟的发动机)和低工况性能的对比见表2。

表2　WR-21两种工况下性能对比

| 性　能 | 100%工况 | 30%工况 |
|---|---|---|
| 功率/MW | 24.86 | 9.19 |
| 耗油率/(g/(kW·h)) | 203 | 208 |
| 动力涡轮入口温度/℃ | 852 | 852 |
| 排气温度/℃ | 355 | 272 |
| 压比 | 16.2 | 8.1 |
| 空气流量/(kg/s) | 73.2 | 39.1 |
| 输出轴转速/(r/min) | 3 600 | 2 600 |
| 质量/kg | 45 975 | |
| 箱装体长/mm | 8 080 | |
| 箱装体高/mm | 4 830 | |

此外，美国海军从便于舰艇现代化改装出发，一开始就规定WR-21的尺寸不能大于LM2500箱装体，两者的"脚印"要相同，以便换装。

WR-21借增加中间冷却器、回热器和涡轮可变几何导叶，使与典型的美国海军燃气轮机相比明显改进了性能。按通常的概念，硬件的增加会降低可靠性，增加维护和支援费用，使维护更困难。为了解决这一问题，设计时应始终对可能性坚持高标准要求。

WR-21采用单元体设计思想，共分16个单元体，分别为低压压气机、中介机匣、高压核心支承、高压涡轮导向器、高压涡轮、动力涡轮进口可变几何导叶、动力涡轮、动力涡轮支承结构、管系、附件传动、中间冷却器组件、控制系统和回热器。

## 四、部件特点

### (一)压气机

高、低压压气机均由 RR 公司的 PB211-535 航空发动机派生而来。低压压气机为焊接式钛鼓转子,上有六级钛合金叶片,静叶则装在铝合金的可中分机匣上,机匣上带有放气阀。高压压气机也是六级,压比为 4.9。

### (二)中间冷却器

中间冷却器是一个双回路系统:一个是在发动机上的淡水-乙二醇(两者体积比为 1∶1)闭式中间回路与空气的热交换系统,其是由 5 个铜-镍鳍片板式逆流热交换器组成,冷却剂流量为 3 400 L/min;另一个是在机外的淡水/海水热交换器,海水流量 5 300 L/min,由船上的海水循环系统提供。控制装置可根据入口空气的温度和相对湿度,及低压压气机出口空气压力自动控制冷却剂绕过海水热交换器的旁通阀,以防止在高压压气机中产生冷凝。

### (三)燃烧室

燃烧室(图 3)由 9 个径流式火焰筒组成。高压压气机排出的空气经排气总管到回热器进口,从回热器来的空气经总管进入燃烧室。该径流式结构与干式低排放燃烧室发展计划相容,可允许方便地翻新改进。每个火焰筒均可在箱装体内部单独拆换。

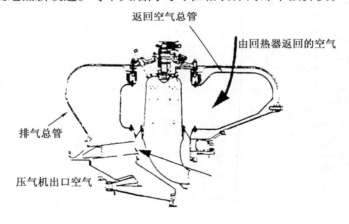

**图 3　燃烧室**

### (四)涡轮

高压涡轮由 RB211-524 航空发动机的高压涡轮派生,其特点为单级、轴注,导叶、动叶均为气膜冷却叶片。为降低涡轮的通流能力对叶片进行了修改。

低压涡轮由 RB211-535 的低压涡轮派生,单级轴流涡轮,为增加涡轮的通流能力,对叶型进行了修改。

动力涡轮入口可变几何导叶在 40% 功率时全关(最小流量位置),在 100% 功率时全开,54 片导叶由齿环带动,齿环则是由 WR – 21 发动机的控制器控制的两个液压作为动器驱动,每个导叶均可单独更换。

动力涡轮共五级,是基于 Trent700 和 Trent800 发动机设计的新设计。动力涡轮在100%功率时的转速为 3 600 r/min。

(五)箱装体

WR – 21 箱装体为弹性安装。箱装体也提供了与船上各供应系统的接口,如电、启动空气等。箱体表面温度的设计值不大于 52 ℃。对于海军型,箱体需可防核、生物和化学污染并隔声、隔热。

箱体的侧壁板和垂直构架在维护需要时可拆卸,当大规模维修时,发动机全长上都是可接近的。在箱体侧壁上有标准的海军门,船员可入内进行日常和预防性维护。底板上有附加的板,分别供辅助齿轮箱维护和通往发动机后支承及排气室用。

辅助齿轮箱安装在燃气发生器下,上有启动机、燃油泵、滑油泵、通气装置。所有附件可从箱体侧面拆卸。此外直接在辅助齿轮箱下的底板上增加了一个通道,这样,燃油泵、滤器、启动机(与 RR 公司 Trent 发动机的启动机相同)注油口、磁力捕屑器和盘动高压转子的插口均可从侧面开口或底板开口,方便接近。

主要部件的拆换通过进气道完成。为此,箱体前部的侧板和顶板是可拆的,箱体内也安装了发动机拆卸导轨、回热器拆卸导轨的结构物安装界面。

在进气喇叭口的上方,装有一个外物防护网,当要拆除发动机时,先将装于箱体顶部的外物防护网从进气道中吊出。因为箱装体是以某一设定的倾角弹性安装在基座上的,因此装有倾角挡块,以降低在任何时候可能发生的箱装体"下蹲"而导致的发动机与齿轮箱间的不对中量。倾角挡块是两个附加的,各可承受 2 270 千克力[①]的支座,轴向地设置在箱装体的端面,使箱体免于倾角增加。除倾角挡块外,在箱装体的角上还装有销、环系统。测量并记录箱装体初始安装时销相对于环的值,当出现任何测量数据的差别时,都意味着发动机相对于船体结构发生了移动。倾角挡块和销、环系统预期可减轻美国海军在现有发动机上反复进行的计划维护的工作量(现有发动机是在高速连接轴上测量偏斜和平行偏移)。

(六)回热器

回热器是板翅式逆流热交换器,采用 14 – 4 不锈钢。回热器可以正常方式运行,也可旁通运行。旁通时空气从回热器的进气管直接进入出气管返回燃烧室,而不经过回热器。发动机可以旁通方式在全工况运行,但是耗油率高。

当旁通运行时,回热器模块的温度升高,因为空气侧的流量为零。这种较高温度的运行模式用于回热器短期运行后的清洁。除了短期运行后的清洁(在旁通方式下是自动完成的),回热器的维护仅限于目视检查。

---

① 1 千克力 =9.806 65 N(准确值)。

## (七)控制系统

WR-21 控制系统是基于开式结构 Futurebus+母板的全功能控制器。控制系统对发动机进行控制和监测,并提供机旁操纵台的接口,必要时就地操纵和诊断发动机。该系统在所有关键的运行控制功能上都有充分的冗余度,控制装置位于一个 1 520 mm × 610 mm × 610 mm 的环境密封的柜内。控制器的主要功能是程序启停、稳态和瞬态控制、监视、故障探测和超速、振动超限、滑油压力过低、涡轮进口温度过高等状况下的应急停机保护。控制器也对涡轮可变几何导叶以及中间冷却器和回热器的旁通进行控制。

控制器有自诊断能力。故障诊断降至船上维修级,失效元件可由船员及维修人员更换。控制器也能隔离失效的系统传感器。

正常运行时,控制器监测遍布 WR-21 各系统的数百个温度和压力,确定发动机的状态,做出控制发动机性能的逻辑决定。如有需要,还可向用户提供最新的数字健康诊断系统和低价位的自动记录装置。

## 五、其他维护

### (一)孔探能力

WR-21 孔探能力综合如表 3 所示。

表 3 WR-21 孔探能力

| 部 件 | 孔探位置 |
| --- | --- |
| 低压压气机 | 在每级的左、右舷侧 |
| 高压压气机 | 在每级的左、右舷侧 |
| 高压涡轮 | 在相对于燃烧室火焰筒的 9 个位置 |
| 低压涡轮 | 在相对于燃烧室火焰筒的 9 个位置 |
| 动力涡轮 | 在每级的左、右舷侧 |
| 涡轮可变几何导叶 | 54 个导叶中的每一个 |
| 燃烧室 | 每个火焰筒、燃气收集器 |
| 中间冷却器芯 | 每个芯通过水洗喷嘴孔进行孔探 |

### (二)流道清洗

通流部分流道用水洗喷嘴清洗。因为 WR-21 装有中间冷却器,因而水洗喷嘴装在两处,装在喇叭口处的前 20 个喷嘴向低压压气机的内下方喷射,清洗低压压气机。另一组喷嘴装在中间冷却器的外壳上,雾化颗粒很小,覆盖整个热交换器的外表面,在液体流过热交换器后继续清洗高压压气机。第二级喷嘴由 40 根小而均布的水洗总管组成,每根总管上装有 3 个细雾化喷嘴。

WR-21水洗系统可以离线清洗也可在线清洗,可用美国海军批准的清洗剂来清洗发动机,也可指定使用与DDG-51级驱逐舰上其他燃气轮机曾用的同等质量的漂洗水。

（三）计划维护

在美国海军内部,计划维护是指预防性维护系统。已要求WR-21将计划维护量降至最低,以减少船员维护成本。目前美国海军DDG-51驱逐舰维护量约4.4h/周。WR-21的计划维护综合如表4所示。

**表4　WR-21计划维护表**

| 项　目 | 频　度 |
| --- | --- |
| 燃烧室孔探 | 半年 |
| 低压涡轮孔探 | 半年 |
| 高压压气机孔探 | 半年 |
| 高压涡轮孔探 | 半年 |
| 中间冷却器孔探 | 半年 |
| 低压压气机孔探 | 半年 |
| 动力涡轮孔探 | 半年 |
| 涡轮变几何导叶孔探 | 半年 |
| 更换燃油滤 | 半年+视情 |
| 附件传动连接检查 | 季 |
| 喇叭口密封检查 | 月 |
| 通气器检查 | 年 |
| 排气管支承检查 | 半年 |
| 泄放系统检查 | 年 |
| 电缆接头检查 | 年 |
| 发动机与中间冷却器的结合部检查 | 日 |
| 发动机支架检查 | 年 |
| 排气室密封检查 | 季 |
| 火焰探测器 | 季 |
| 异物防护网 | 半年 |
| 燃油管与发动机的连接 | 日 |
| 齿轮箱连接检查 | 月 |
| 热探测 | 半年 |
| 高压压气机放气系统 | 月 |

表 4(续)

| 项　目 | 频　度 |
| --- | --- |
| 温度传感器 | 季 |
| 点火器导线 | 周 |
| 低压压气机放气系统 | 月 |
| 中间冷却器旁通 | 月 |
| 滑油滤器 | 月 |
| 滑油油位 | 周 |
| 磁力捕屑器 | 周 |
| 回热器管道 | 半年 |
| 轴对中检查 | 半年 |
| 启动机油 | 月 |
| 通气挡板 | 半年 |
| 清洗雾化状况 | 季 |
| 系统漏泄检查 | 周 |
| 清洗发动机 | 24~48 运行小时 |
| 中间冷却器乙二醇取样 | 周 |

## 六、结束语

21 世纪,高性能水面战舰将采用综合全电力推进(IFEP)系统,美、英、法等国海军都制定了相应的 IFEP 系统发展计划。美国海军海上系统指挥部预计,如果采用 ICR 燃气轮机可使总运行油耗下降 23%~25%,如果采用综合电力系统可使总运行油耗降低 15%~19%,这两种技术的结合可使油耗降低 36%~38%。WR-21 燃气轮机正被考虑作为综合全电力推进系统的主要设备。

## 参 考 文 献

[1] WELLER C L, BROADBELT A, LAW B. WR-21 design and maintenance[C]//ASME 1996 International Gas Turbine and Aeroengine Congress and Exhibition. American Society of Mechanical Engineers Digital Collection, 1996.

# 舰船燃气和蒸汽动力装置的发展与展望*

闻雪友

(哈尔滨船舶锅炉涡轮机研究所)

**摘 要**:通过对两个典型机型(WR-21,GT25000)及其他有关资料的介绍,综述了舰船燃气及蒸汽动力装置的现状与发展趋势。

## 一、WR-21——新一代的船用燃气轮机

近年来,船舶及工业燃气轮机领域的人们总是情不自禁地把关注的目光投向 WR-21 燃气轮机的研制工作。WR-21 是一型带有中间冷却器和回热器的燃气轮机。与简单循环相比(如 LM2500),其不仅输出功率增加,额定功率下的耗油率降低,而且在大部分功率范围内具有平坦的耗油率曲线,同时又具有低噪声及低的排气红外特征,与美国海军现役的燃气轮机推进动力水面舰船相比,采用该动力装置的舰船年燃油消耗量降低约 30%。该机现已成为新一代船用燃气轮机的象征。

约 50 年前,船用燃气轮机的先驱——英国认识到船用燃气轮机要求在整个功率范围内有低的耗油(尤其是部分负荷),为实现这一点,研究人员尝试选择了包括 3 台压气机、2 台中间冷却器、1 台回热器、3 台涡轮的复杂循环,即 RM60 燃气轮机。该型机于 1951 年进行了陆上试验,功率达到 3 895 kW(5 300 hp)。作为船用燃气轮机,这个装置还是值得注意的,因为尽管该装置复杂,但其比重仅达到了 3.59 kg/kW(2.64 kg/hp),在 40%~80%功率范围内的耗油率曲线平坦。1954 年该机装于"灰鹅"号炮艇,海上航行证实了装置的可靠性。RM60 虽然达到了良好的部分负荷性能,但却使发动机的购置成本大大提高,且其体积庞大、控制复杂,循环系统的损失也较大,加之不久其性能即被改进的简单循环发动机("海神")所超越,因而未获得进一步发展。

几十年过去,情况变得不一样,事物螺旋上升的规律在这里得到体现。由于燃气轮机

---

\* 文章发表时间:1999 年 4 月。

及热交换器方面的技术进展，使得间冷－回热循环在达到最高的循环效率、优良的变工况性能的同时仍能使装置结构紧凑，这使其可能特别适于舰用。

1981 年，Rolls－Royce 公司向美国海军提出发展间冷－回热燃气轮机（ICR）的设想。

1985 年秋，美国海军首次招标 38 ℃，17 897 kW 的 ICR 高效舰用燃气轮机。

1987 年底，美海军决定将该燃气轮机的功率提高到 19 686 kW，并考虑采用全电力推进方案。到 1989 年底，美海军也只是维持 GE 及 Rolls－Royce 公司的预研工作。

1990 年，美海军恢复 ICR 计划，重新招标，确认功率为 19 686 kW 并有 10% 的增长潜力，尺寸适合于 LM2500 换装。Rolls－Royce 公司与 Westing house 公司合作，结合 RB21 系列航空发动机单元体结构，获得了与航空型零件高度通用的最佳 ICR 方案。

1991 年 12 月，美国海军将 ICR 燃气轮机机组的设计和发展合同授予 Westinghouse Electric Coporation（WEC）船舶分部，分承包商主要有 Rolls-Royce 公司工业与船用燃气轮机分部（负责燃气轮机），Allied Signal 公司航空系统和设备集团（负责回热器和中间冷却器），CAE 电子公司（负责控制设备）。

1991—1995 年为前期发展阶段（设计、发展、调试和确认主要性能参数）。

1994—1999 年为全尺寸发展阶段（美国海军试验、调试、训练、编写说明等）。

1997 年的 5—9 月，由 Northrop Grumman 船用系统公司（新的美国总承包商）与 RR 公司共同在英国国防评估和研究机构的 Pyestock 试验场完成了产品标准系统的最新 500 h 耐久性试验，共试验 503 h，69 次启动，相当于海上实际运行 1 500～2 000 h。计划于 1998 年美国海军在 Philadephia 的海军水面舰船中心的 Carderock 分部建的第二个试验场上进行第二台机的试验。

原计划在 1999 财政年度分批装备 DDG－51"Arliegh Burke"级导弹驱逐舰。但按美国海军的计划，第一台间冷、回热燃气轮机要到 2004 财政年度才装备于战舰，比预计的晚 5 年，装备最后 9 艘 DDG－51 级舰。

从 1994 年 7 月首台前期发展型 WR－21 开始试验以来，多台 WR－21 试验型、发展型及生产型样机已相继完成各项性能试验和耐久性试验。至 1997 年，总试验时数已超过 1 200 h，其中包括 110% 负荷的重要试验。在首台样机试验前完成了各模块和部件试验。预计 3 000 h 验收试验将在 1999 年进行。WR－21 表现出的高性能，已引起世界各国海军的极大兴趣，目前考虑采用此新型动力装置的舰船有：美国海军 DD－21 级驱逐舰、DDG－51 级驱逐舰，英国新型航母（以替代 2010 年退役的 3 艘"无敌"级轻型航母），英国、法国、意大利"地平线计划"中的新一代"前景"级护卫舰（CNGF）。

ICR 循环与简单循环燃气轮机的不同之处在于：

（1）在 2 台压气机之间设有 1 台中间冷却器，使得空气在进入高压压气机前被冷却，从而减少高压压气机所需的功耗，提高高压轴的效率并增加发动机的功率（在 WR－21 上约增加 25% 的功率）。此外，间冷也降低了高压压气机出口的空气温度，提高了空气与燃气排气的温差，从而增加了回热器的效率。

（2）高压压气机出口的空气先进入回热器，利用燃气排气余热对空气进行预热，然后进入燃烧室，这就降低了为达到预定的燃气初温所需的燃料量，提供了在耗油率方面的重要改进。

（3）从燃气发生器出来的燃气，通过第一级可变几何导叶（VAN）进入动力涡轮。随着功率减小，动力涡轮质量流量减少，可变几何导叶逐渐关小。因此，对一给定的部分负荷可保持高的燃气初温（在 WR-21 中，全工况和 30% 工况下动力涡轮入口燃气温度均为 852 ℃，出口温度则从全工况的 355 ℃ 降为 30% 工况时 272 ℃），这使回热器空气-燃气的温差增加，改进了回热器的传热，使部分负荷下的耗油率获得改善。在 WR-21 上采用回热加 VAN 可节省燃油 30%~40%。

回热器、中间冷却器和涡轮可变几何导叶是使 WR-21 相对于简单循环获得耗油率减小和功率增大的三个关键特征。

WR-21 在 ISO 条件下的全工况性能（对于发展成熟的发动机）和低工况性能的对比见表 1。

表 1　WR-21 两种工况下性能对比

| 性　能 | 100% 工况 | 30% 工况 |
|---|---|---|
| 功率/kW | 24 843（33 800 hp） | 9 188（12 500 hp） |
| 耗油率/(g/kW·h) | 204（150 g/(hp·h)） | 208（153 g/(hp·h)） |
| 动力涡轮入口温度/℃ | 852 | 852 |
| 排气温度/℃ | 355 | 272 |
| 压比 | 16.2 | 8.1 |
| 空气流量 | 73.2 | 39.1 |
| 输出轴转速/(r/min) | 3 600 | 2 600 |
| 质量/kg | 45 975 ||
| 箱装体长/mm | 8 080 ||
| 箱装体高/mm | 4 830 ||

21 世纪高性能水面战舰将采用综合全电力推进（IFEP）系统，美、英、法等国海军都制定了相应的 IFEP 系统发展计划。美国海军海上系统指挥部预计：如果采用 ICR 燃气轮机可使总运行油耗下降 23%~25%；如果采用综合电力系统可使总运行油耗降低 15%~19%。这两种技术的结合可使油耗降低 36%~38%。WR-21 燃气轮机正被考虑作为综合全电力推进系统的主要设备。

## 二、GT25000 船用燃气轮机

苏联海军的燃气轮机舰船占世界第一位，20 世纪 60 年代以来共有巡洋舰 8 艘、驱逐舰 41 艘、护卫舰 68 艘、快艇 94 艘、气垫船 36 艘。苏联船用燃气轮机发展走专门船用设计的道路。

除苏联最初的船用燃气轮机装置 M-1(功率为 2.9 MW(4 000 hp),寿命为 100 h,耗油率为 57.8 g/(kW·h))是在航空发动机的基础上发展而来的,其后从 M-2 燃气轮机装置开始都是独立研制的。该机寿命为 1 000 h,功率为 11 MW(15 000 hp),耗油率为 353.7 g/(kW·h)(260 g/(hp·h)),是苏联第一台全工况船用燃气轮机。

1958 年,为"乌克兰共青团员"号舰(北大西洋公约组织称之为"卡辛"级)研制了 M-3 主动力装置,功率为 26.5 MW(36 000 hp),由两台燃机组成,寿命 1 000 h,耗油率 353.7 g/(kW·h)(260 g/(hp·h))。该舰的特色是采用全燃推进,有可正、倒车的减速器,当时(1962 年)世界上还没有。英国类似的装置在 1969 年才用于 21 型和 42 型,美国类似的装置则直到 1973 年才用于驱逐舰"斯普鲁恩斯"号上。

1965—1966 年起,开始研制第二代发动机,主要任务是提高经济性、寿命,改善声学特性。研究结果表明,舰船主动力装置应由不同功率的发动机通过连接系统组成一整体,使其在任何航行工况下都能经济的工作。用这样的概念研制成 M-5,M-6,M-7 全燃推进装置,其中包括独立的、不同功率的巡航和加速机。

所采用的系统可保证在舰艇全速时全部装置投入工作,也可仅一台发动机工作,通过减速器将功率分配在两只螺旋桨上。在这些装置中有世界首次采用的倒车动力涡轮、快速作用的气动离合器等一系列新技术。1971 年开始研制第三代燃气轮机为 ГТД3000、ГТД15000。

ГТД3000 为三轴燃气轮机,采用简单循环,功率为 3 000 kW,用作快艇和动力效应船的主动力,已累计运行 55 000 h。

ГТД8000 也采用三轴、简单循环,功率为 6 000~8 000 kW,可用作快艇的加速机组,也可用于军、民用动力效应船,同时也是"光荣"号巡洋舰上的巡航发动机。其船用型发动机已累计运行 $13 \times 10^4$ h。

ГТД5000 也采用三轴、简单循环,功率为 15 000~17 000 kW,用作各种排水型水面舰艇的全工况或加速机组。与第二代燃气轮机相比,其燃气初温提高了 200~250 ℃,压比提高了近一倍。而且由于在高、低压涡轮上均采用了高应力单级涡轮代替两级涡轮,转子采用双支承,以及高性能冷却叶片、新材料、新工艺的应用,使发动机的比重仍可接近航空发动机的水平。基于这些发动机,对气垫船、水翼艇、水面舰艇设计研制了不同的装置,其中最有代表性的是安装在排水量为 13 000 t 的轻巡洋舰"光荣"号上的一套燃-蒸联合循环装置(COGAS)。该装置作为巡航机,可以提高巡航时的经济性,并以大功率燃气轮机作为该舰的加速机。

从 1986 年起,开始研究新的大功率船用燃气轮机 GT25000。

GT25000 是苏联舰船燃气轮机研制单位——现乌克兰"机械设计科研生产联合体"新研制的第四代燃气轮机。GT25000 的主要技术性能如下:

(1) ISO 的主要技术性能

①动力涡轮输出法兰处的功率为 28 670 kW(39 000 hp);

②在 28 670 kW 下的效率为 36%。

(2) 发动机的尺寸

①从进气管前端至弹性轴输出法兰的长度不大于 8 200 mm;

②宽度不大于 2 500 mm;

③从底架安装面至排气管罩壳顶部的高度不大于 3 500 mm。

(3) 发动机质量

①发动机本体质量 12 t;

②包括底架、排气管、燃气罩壳、排气管罩壳在内的发动机质量不大于 21 t。

(4) 发动机大修前寿命 25 000 h。

(5) 大修前发动机服务年限 10 a。

GT25000 燃气轮机适于驱逐舰、护卫舰用作主推进动力。

## 三、大功率蒸汽动力装置

世界上各主要海军国家都曾为舰船蒸汽动力装置的发展和应用倾注了巨大的人力、财力和物力,使蒸汽动力装置在大中型水面舰船上获得广泛的应用。美国在第二次世界大战后已研制发展了两级 8 艘常规动力的大型攻击型航母,研制了五代常规动力的大型驱逐舰和两代中型驱逐舰。俄罗斯(含苏联)在战后也研制发展了两级 6 艘常规动力的航母,研制了五代大型驱逐舰和七代中型驱逐舰。这些舰船大部分采用蒸汽动力装置,少部分采用燃气轮机装置和柴燃联合动力装置,而大中型常规动力航母则全部采用蒸汽动力装置。

在世界各国的大中型水面舰船上,蒸汽动力装置通常采用 2 台主锅炉配 1 台主汽轮机组带动 1 根螺旋桨轴的形式。驱逐舰采用双轴推进,航母则采用四轴推进。装置中的主动力设备可达到一定程度的兼容。

现今,国外舰船蒸汽动力装置达到的水平如下:

(1) 每台主蒸汽锅炉的容量为 65 ~ 120 t/h,每台主汽轮机组的功率为 27 500 kW(35 000 hp)、36 800 kW(50 000 hp)和 51 500 kW(70 000 hp)。

(2) 采用的蒸汽初、终参数为 4.4 ~ 6.3 MPa,450 ~ 490 ℃,9.8 ~ 19.6 kPa。美国航母则采用 8.27 MPa 和 510 ℃ 的蒸汽初参数。

(3) 蒸汽动力装置全航速时耗油率为 381 ~ 558 g/(kW·h)(280 ~ 410 g/(hp·h))。必须指出,蒸汽动力装置耗油率中包含了全舰性主要能源的消耗(这有别于其他动力装置),因而在评价各种动力装置经济性时应当有所区分。

(4) 采用蒸汽动力装置的舰船的续航力,在 18 ~ 20 kn[①] 巡航工况时,驱逐舰级为 3 200 ~

---

① 1 kn = 0.514 44 m/s。

6 000 n mile①；对于航母，则可达 8 000 n mile。

（5）蒸汽动力装置的比重为 12～20 kg/kW（9～15 kg/hp）。

（6）蒸汽动力装置的控制、保护和热工监测已逐步采用二级或三级集散式自动控制系统，已从就地控制、监测走向集控室、动力部位和舰桥等部位的强控和自控。

我国舰船蒸汽动力装置是从无到有逐步发展起来的，经历了使用、研仿和自行研制等过程。目前，我国有单轴推进功率 11.756～26.500 kW（16 000～36 000 hp）的多型蒸汽动力装置，并从设计研究、生产制造、调试投运到服役中保养维修等已都完全立足国内。这些独立研制的主动力装备为现今海军主力舰船的主动力提供了重要的技术和物质支撑。新一代主锅炉的热效率比前一代产品提高十个百分点，主机组提高了三个多百分点，明显地降低了动力装置的耗油率，提高了舰船的续航力。蒸汽动力装置采用了全电集散式二级调控、监测系统，使蒸汽动力装置的自动化水平有了明显提高。

舰船蒸汽动力装置拥有许多主、辅动力装置和数量可观的工作系统，是比较复杂的系统工程。发展和研制一型蒸汽动力装置需要花费相当数量的人力、财力和物力，还需要较长的工作周期，而舰船对于动力的需求又是各不相同的。我国作为发展中国家就必须发展兼容性好的蒸汽动力装备和工作系统，利用合理的组合满足对动力的需求。

舰船蒸汽动力装置是"可靠、顶用、经济、立足国内"的舰船主动力，有明确的需求和良好的发展基础，只要坚持持续稳定、协调发展的舰船动力技术政策，只要勤奋务实地工作，就一定能实现既定的发展战略。

## 参 考 文 献

[1] WELLER C L, BROADBELT A, LAW B. WR-21 design and maintenance[C]//ASME 1996 International Gas Turbine and Aeroengine Congress and Exhibition. American Society of Mechanical Engineers Digital Collection, 1996.

---

① 1n mile = 18 582 m

# 双工质平行－复合循环热机
# （程氏循环热机）I *

闻雪友

**摘　要**：双工质平行－复合循环热机是一种较新的热机，其能达到高效率和高比功。文中对该循环进行综述，涉及循环原理、循环特点、循环分析、各循环比较和应用实例。

## 一、引言

在燃气轮机中以喷注蒸汽方式改进效率，增大功率，这种方法并不是全新的。近年来，由于对节省燃料和降低投资成本的要求日益突出，因此，这种方法重新引起研究人员的兴趣。

国际动力技术公司的程大猷先生于1976—1981年间提出"双工质平行－复合循环热机"专利发明，又称程氏循环。

程氏循环发动机有两种分离的工质，每一工质分别增压，然后以一种简单的方式混合、膨胀和回热。该循环本质上是平行联合：一个布拉东循环和一个回热朗肯循环系统，以布拉东循环之压比作为运行极限，以朗肯循环之温度为上限，并对两个循环之排气余热均加以利用。采用朗肯循环工质回热是该循环的另一个非常重要的特点。

应当指出，一个发动机同时用两种工质运行并不是全新的概念。程氏循环专利发明的独特之处在于通过适当选择循环参数或部件的独特匹配而达到高效率、高比功的目的，以及确定双工质平行－复合循环的运行极限。

## 二、双工质平行－复合循环热机原理

典型的双工质平行－复合循环热机的原理图如图1所示。空气经一调节空气压力的节流阀1进入压气机2。如果压气机压比低于12，入口空气温度27 ℃，则节流阀1也可作为一个汽化器，一部分燃料可从此节流阀中加入，如图中4′所示。如果压气机压比大于12，又没有专门的冷却，那么在气－燃料混合物被压缩时就会自燃。故而，对于高压比，燃料必

---

\* 文章发表时间：1986年4月。

须在压气机后4处加入。

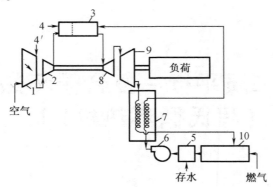

**图1　典型的双工质平行－复合循环示意图**

图1中3为燃烧室,除了燃烧外,第一种工质也可以用其他方式加热,如太阳能、核能等。

第二种工质(例如水)用泵6压到高压,进入热交换器7,从动力涡轮9中排出的排气(蒸汽/燃烧产物的混合物)废热中吸收热量,水被加热成蒸汽。多数情况下是过热蒸汽,湿蒸汽也是可能的。如果从热交换器出口是热水或蒸汽/水的混合物,则在进入燃烧室3后亦迅速蒸发成过热蒸汽,通过两种工质的紊流混合,热能从热燃烧产物输入到蒸汽。蒸汽与燃烧产物的混合是在燃烧后进行的,用蒸汽来控制燃烧产物的温度,达到设计的涡轮进口温度值。

然后,双工质混合物进入驱动压气机的涡轮8和动力涡轮9。从动力涡轮出口的排气进入逆流热交换器7。热交换器的气侧为混合气,其温度从动力涡轮出口的排气温度降至略高于混合气中的水饱和温度的程度。热交换器的液侧,水在压力下从环境温度加热到沸点并蒸发,在水/蒸汽混合区形成湿蒸汽,如果能从排气中吸收足够的热量,那么热交换器出口就是过热蒸汽。

从热交换器出来的混合气在开式循环中就排入大气。在闭式循环中则进入冷凝器10,水从顶部喷头喷淋而下,水滴从混合气中吸收热量,混合气中的水凝结,同冷却水一起落入容器底部回收,气体容器顶部通入大气。

水进行适当处理后泵送到热交换器的液侧,一部分水通往冷却塔或其他冷却器,然后在冷凝器中重新回用。

由于循环中有两种工质,水、空气/燃料燃烧产物。每一种工质分别增压,继之以简单的混合、膨胀、回热。该循环本质上是平行联合,即一个布拉东循环和一个回热朗肯循环系统,由于两种工质混合,故输出又叠加在一起,谓之复合。该循环实质上是一种双工质平行－复合回热热机。

由图2可进一步说明。该图未考虑损失,而且为解释方便,两种工质在其各自的布拉东和朗肯循环中分别处理。前已述及,实际上两种工质在循环的一定阶段是平行耦合的,因此虽然图 2 中用两个单独的循环 $T-S$ 图来表示,然而两者却是关联甚密的。

气体工质从状态 1 始经压缩达 2,燃烧及与蒸汽混合使热力状态达 3,然后与蒸汽一起膨胀到 4,排气热量传给另一工质并冷却,理论上回到热力状态 1。

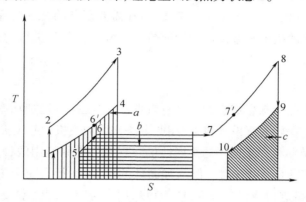

**图 2　循环温熵图($T-S$ 图)**

图 2 中,液体工质在 5 泵略高于 2 的压力,温、熵的变化很小。高压液体吸收涡轮排气的热能达到沸点温度 $T_6(T_6 < T_6')$,液体在达到饱和点 7 以前被连续加热,甚或加热到略低于 $T_4$ 的过热温度 $T_7'$,蒸汽与空气/燃料燃烧产物混合达 $T_8(T_8 = T_3)$,并一起膨胀到 $T_9(T_9 = T_4)$。蒸汽的排气热量沿 9→10 的通路传给进入的液体/蒸汽,构成一个独特的回热蒸汽循环。排气中的蒸汽则从混合气中凝结出来,返回热力状态 5。

## 三、与其他循环的区别

程式循环双工质热机初看起来与燃-蒸联合循环、回热燃气轮机循环、喷水燃气轮机等循环有些类同,因此有必要对几种循环的发动机的差别加以讨论。

### (一)燃-蒸联合循环动力装置

联合循环仅是在观念上将布拉东和朗肯循环结成一体。两个循环串联运行,布拉东循环结成一体。布拉东循环的废热用作热源,使在一分离的朗肯循环中的水沸腾,两种工质也没有发生混合,输出功也是由各循环各自的涡轮产生的。

联合循环中各工质有各自的环路,有两个各自独立的动力涡轮;而程氏循环是两种工质混合,仅需一个动力涡轮。联合循环中的锅炉非常类似于程氏循环中的回热热交换器,两者在热端和颈部均经受同样的温度限制。然而,在联合循环中对热交换器的限制更大,因为联合循环中的水必须加热到足以使蒸汽在锅炉出口处进入过热蒸汽区。程氏循环则无此要求,因为即使热交换器出口是湿蒸汽,也将在燃烧室中变为过热蒸汽。当然,根据高效率或高比功的要求,对双工质平行-复合循环发动机也有不同的制约。

### (二)回热燃气轮机

双工质循环与回热燃气轮机循环在感觉上的类同性在于两者似乎都是从动力涡轮后

将废热回收到循环中。但是,在回热燃气轮机中仅有一种工质,回热是将压气机排气在进入燃烧室前完成预热。由于涡轮排气温度必须高于压气机出口温度,故而回热对发动机的压比有制约,在高压比的布拉东循环中采用回热,其收效减小。

此外,在回热燃气轮机中,虽说是单工质,但却需要复杂的往返管道来实现回热。在双工质循环中没有这样的机械复杂性。因为回热是在两种工质间进行的,热交换器易于布置在动力涡轮排气处。

(三)喷水燃气轮机

燃气轮机中喷水是冷却燃烧室的一种有效方法,也是一种增加功率或推力的较简单的方法。在早期的航空涡轮喷气发动机上,用此方法来增加飞机的起飞推力。

喷水燃气轮机循环与程氏循环的类同性仅表现在涡轮中均用两种工质。然而,虽然两者都用同样的两种工质(水和空气),但两种循环的设计、运行是完全不同的。在喷水燃气轮机中,水可从压气机进口或出口喷入,或为冷却用直接喷入燃烧室,水没有从循环中回收任何废热。由于水具有大的蒸发潜热,故作为冷却介质特别有效,但是因为没有回热,故此过程对热效率有不良的或几乎很小的影响。

喷水的另一目的是提供短时推力或功率增量,发动机不是为带水连续运行设计的,故而可能加到循环中的水量受压气机失速特性的限制。

双工质平行-复合循环发动机则是设计成以加注蒸汽做连续运行。在双工质循环中,朗肯循环的流体是工质,而不是冷却剂。为达到双工质平行-复合循环发动机的高热效率,循环参数的适当组合将要求汽气比随设计点涡轮进口温度的增加而增加,而在早先的喷水燃气轮机设计中,涡轮进口温度增加总是导致水气比降低。

(四)带喷水和回热器的燃气轮机

在燃气轮机中喷水的一个更近的应用是大气污染控制。水在压气机后喷入空气流中,达饱和点。如果循环中用回热器,则水在热交换器入口以能使水完全蒸发的适当量(小于8%)喷入,气-汽混合物在进入燃烧室前回收排气热量。蒸汽的作用是稀释空气,使燃烧室中的火焰温度降低。燃气轮机$NO_x$的生成与燃烧区内的当地温度有很大的关联,喷水的结果是降低了$NO_x$的水平。

蒸汽的比热容大致比空气大一倍,水的比热容又约是蒸汽的两倍。由此,以水进行热回收(程氏循环即如此)比以汽-气混合物进行热回收更有效。此外,对于一个用气-汽回热系统的回热燃气轮机,同样具有压比限制,最佳压比一般是 6~10。虽然该循环能增加输出,效率也略有改善,但其在输出和效率上的收益远小于程氏循环。

(五)喷注蒸汽的涡轮热机

喷注蒸汽的燃气轮机在过去已有过尝试,但实际上效率相当低,相反一系列联合循环发动机更有吸引力并已付诸商业使用。

这一事实表明,还没有把这样的循环看作两个独立热力循环的连接,发动机设计参数和两个循环的结合是独特的。发动机的运行参数互相关联,对最大比功或效率来说是限定在一个狭窄的运行范围内。例如,蒸汽太多会使蒸汽循环很差,因为其没有纯蒸汽循环的高压比,若蒸汽太少,则发动机与回热燃气轮机差别很少。

早期的分析表明,人们还没有充分认识到参数间这种相互依存关系,或者是没有找到发动机最佳参数的狭窄范围,仅在此范围内发动机效率达到最大。

## 四、程氏循环特点

(一)高热效率

1. 发动机在两工质间折中选择压缩功和蒸发潜热,以最小的总功和热耗达到高压、高温、高焓的涡轮进口条件。与燃气轮机中压气机压缩空气的功耗相比,水加压的泵功可忽略。虽然压水功耗不大,但将水变为蒸汽则需大量的能。然而当压力增加时,蒸发潜热就变得较小,例如,在 3 206 psi[①] 时其为零,当然,此时泵水的功不能忽略。与通常的蒸汽轮机相比,蒸汽过热到很高的温度,因此潜热是蒸汽总含能量的一小部分。故而循环沿一实际的低能耗通路运行,达到高压、高能状态,准备膨胀。

2. 在程氏循环中,燃料可以接近理想配比燃烧,这意味着不再像通常燃气轮机的等压过程那样需要过量空气来使工质温度控制在容许范围内,也就是说,可节省压缩过量空气的功耗(尤其对高压比发动机)。

3. 空气-燃料燃烧产物与蒸汽通过紊流混合的方法,直接将燃料燃烧的化学能传递给蒸汽。这种方法要比常规的通过锅炉管道传热的效果更好。

由于水比气有更高的传热系数,因此在热交换器中一侧用水比用气更好。

在闭式循环中,排气的冷凝被作为一种传热给回热工质(水)的方法。水温趋于其沸点并在回热器中保持该值,其间仍然吸收能量,这导致排气温度和回热温度间有较高的温差,从而达到一个较高的传热比。

4. 回热系统是独特的,仅用朗肯循环回热不理想,仅用高压比的布拉东循环发动机回热也不经济。换言之,在程氏循环发动机中,有可能实现高效运行。

5. 因有蒸汽可用,故可用其作为冷却介质来冷却涡轮而无须用压气机抽气,这进一步减小了压气机的功耗。

6. 在低的涡轮进口温度下,循环的优点消失,在 600 ℃ 以下,效率比朗肯循环差。因此,对程氏循环来说,涡轮进口温度越高越有效。

(二)高比功

由于第二工质为水,在等压下的蒸汽比热容至少要比空气-燃料燃烧产物的比热容高一倍。换言之,蒸汽作为主工质能比燃气轮机中的空气-燃料燃烧产物做更大的机械功,这使双工质发动机达到高功率密度。

对应于最大效率或最大比功的汽气比较大,这使发动机在达到高效率时有高比功。

---

① 1psi = 6.894 8 kPa。

## (三) 变工况性能良好

1. 发动机允许压气机和涡轮的质量流量不等,因此变负荷时可用改变蒸汽量的简单方法来控制发动机的输出,保证高压部分以良好的效率运行。

发动机沿一最小能耗通路将两种工质带到高压、高温的运行条件,这种选择能使部分负荷的运行效率较高。

2. 该循环发动机对部件效率影响的敏感性较小,因为其部分回收因循环本身损失而产生的废热。例如,在一个燃气轮机里,当压气机效率从90%降到84%时会使总效率损失超过6%,而在程氏循环里所引起的总效率损失远小于此值。

对程氏循环来说,排气背压的变化也没有引起像通常燃气轮机那样严重的损失。

## (四) 改善排放标准

燃气轮机中$NO_x$的生成与燃烧区内的火焰温度有关,因此喷注蒸汽的结果是降低了$NO_x$的水平,改进了污染特性。

## (五) 独特的参数匹配

程氏循环专利发明的独特技术在于适当选择循环参数或部件间的独特匹配来达到高效率或高比功,以及确定循环的运行极限。

对双工质平行-复合循环来说,有四个主要参数。在燃气轮机循环设计中的两个独立参数是涡轮进口温度和压气机压比,在双工质循环中还有两个参数:热输入率(或空气燃料比)和汽气比。理论上这两个参数的多种组合都是可能的,但是涡轮进口温度设定后,一旦对最大效率或最高比功的特定要求确定后,两者就不能独自规定。实际上,所有的热机循环都不可能设计成既达最高效率又达最大比功。据此,提出了一个狭窄的循环设计参数区,仅在此区域能实现高效率或高效率与高比功间折中的循环。

图3~图6是一些典型的参数曲线。图3表示汽气比与热输入率的关系,两者是在对应于涡轮进口温度的最佳压比下得到的。下边界线是与涡轮进口温度相应的最佳压比下的最高过热度,表示最好效率线相应的汽气比。上边界线表示在给定的涡轮进口温度下大致最大的汽气比值,是一条发动机性能和比功折中选择的推荐线。

图4总括地表示了比功的狭窄范围。此时,对一给定的涡轮进口温度,对应最佳压比下相应峰值热效率的热输入率和汽气比等参数间的内部关联是确定的。上界则是按最大比功限定的。

图5(a)、图5(b)表示了作为涡轮进口温度函数的热输入率(或空气燃料比)和压比的覆盖范围,这些区域构成了有一个标准热交换器和合理的部件效率的程氏循环热机所覆盖的循环参数组合。图中$E-E$是最大效率线,$P-P$线表示在高效率和高比功间的一种折中,超出$E-E$和$P-P$所包括范围的部分,表示将来研究更高效率的部件和更小温度限制的热交换器设计时所可能实现的范围。图5(a)左部的纵坐标相应于循环用碳氢气体燃料,右部的纵坐标用于煤油。

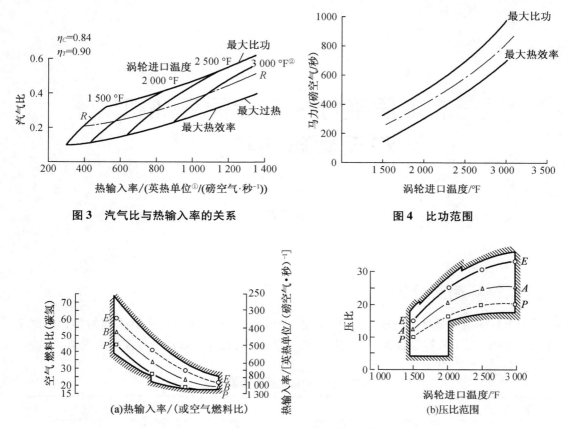

图3　汽气比与热输入率的关系

图4　比功范围

图5　热输入率(或空气燃料比)和压比的范围

图6表示汽气比的范围,表示有合理部件效率的程氏循环发动机循环参数的一个相容组。$E-E$是最大效率线,接近于下边界。高比功$P-P$线大致是上边界,超过$E-E$和$P-P$的区域则是预见到将来发动机部件效率的改进。

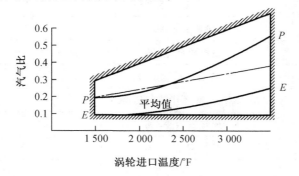

图6　程式循环热机汽气比值范围

① 英热单位,即 Btu,$1\text{Btu} = 1.055\,06 \times 10^3$ J。
② $1\,°\text{F} = \dfrac{9}{5}t + 32$,$t$ 为摄氏温度(单为℃)。

# 双工质平行－复合循环热机
# （程氏循环热机）Ⅱ *

闻雪友

## 一、循环简要分析

为更具体地对循环进行说明，文中引用一个分析结果。该分析中所采用的参数值列于表1。文中研究了某些参数在一定范围内变化的影响，且其他参数一般保持为常值。

表1  分析中所用参数的假定值

| 序 号 | 参 数 | 额定值或范围 |
|---|---|---|
| 1 | 蒸汽质量流量比/% | 0~40 |
| 2 | 涡轮进口温度/K | 1 089~1 644 |
| 3 | 压气机压比 | 6~24 |
| 4 | 压气机多变效率/% | 87 |
| 5 | 涡轮多变效率/% | 87 |
| 6 | 涡轮机械效率/% | 98 |
| 7 | 发电机效率/% | 98 |
| 8 | 压气机进口温度/K | 288 |
| 9 | 余热蒸汽发生器中排气压降/% | 4 |
| 10 | 燃烧室效率/% | 99 |
| 11 | 燃烧室压降/% | 4 |
| 12 | 进口湿度/% | 60 |
| 13 | 燃料热值/(Btu/lb) | 18 600 |

---

\* 文章发表时间：1986年5月。

表1(续)

| 序号 | 参数 | 额定值或范围 |
|---|---|---|
| 14 | 最小温差/℃ | 28 |
| 15 | 余热蒸汽发生器中蒸汽压降/% | 12 |
| 16 | 进口水温/℉ | 288 |
| 17 | 泵效率/% | 70 |
| 18 | 最高金属温度/℉ | 1 089 |

图1表示余热锅炉中最小温差位置的关系(热交换器假定为逆流布置)。对低蒸汽流量(图1(a)),从排气中吸收的热量少,故排气温度线的斜度小,最小温差出现在热交换器的蒸汽出口处,即最小温差等于端部温差。图1(b)表示在较大蒸汽流量下端部温差等于极点温差的情况。此时最小温差同时发生在1,2位置。蒸汽流量进一步增大时(图1(c)),端部温差大于极点温差,最小温差发生在位置2上。在排气进口温度不变时,一旦最小温差的位置移动,锅炉出口蒸汽温度随蒸汽流量增加而降低。

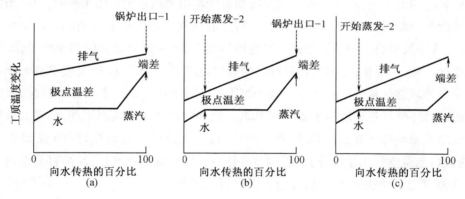

**图1 余热锅炉中工质的温度变化**

图2表示对一典型例子加注蒸汽对循环效率的影响。在锅炉中最小温差位置移位前,效率随蒸汽加入量的增加而增加。在最小温差位置从端部移到锅炉内的极点后,循环效率随蒸汽加入量的增加而减小,因为最小温差位置移动后,蒸汽温度随蒸汽加入量增加而减小。为保持同样的进口温度,需要更多的燃料,此时涡轮流量增加所得的输出净增益不再能平衡燃料的增加,因而效率下降。当然,此时比功继续随蒸汽加入量增加而增加。

在汽气比达0.27时湿蒸汽喷入,在比值达0.35时蒸汽将在锅炉的排气侧开始凝结。0.68的比值表示没有过量空气的汽气比,在此比值下没有过剩空气来维持增加燃料的燃烧,如果涡轮进口温度不允许降低,进一步增加蒸汽量是不可能的。各运行极限的汽气比将随不同的涡轮进口温度和压气机压比而变化。

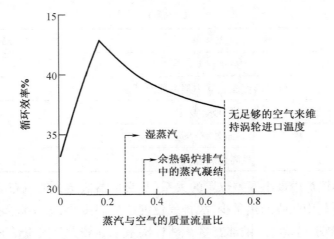

**图2** 循环效率与余热锅炉产生的蒸汽量的关系（涡轮进口温度 1 093 ℃，非冷却涡轮，压比 16）

图3对三个不同的涡轮进口温度表示了循环效率、比功、汽气比和压比间的函数关系。为表明涡轮冷却的影响，对气冷涡轮和无冷却涡轮均由曲线示出。图3表明，对一给定的涡轮进口温度和冷却假定，其最大效率是在一特定的汽气比和压比下达到的。在简单循环燃气轮机中，改进效率的一种手段是提高涡轮进口温度，但为达到最大的效率增益，压比也必须相应增加。然而，需指出的是，在同样涡轮进口温度下，双工质循环相应于最大效率的压比低于简单循环燃气轮机相应于最大效率的压比。对于所讨论图3的五种情况，最大循环效率较简单循环燃气轮机增加了 25% ~ 30%。比功随汽气比增加而增加，与最大效率相应的比功较简单循环燃气轮机增加 50% ~ 99%（按压气进口空气流量计算）或 25% ~ 35%（按涡轮进口当量空气流量计算）。低的百分增量相应于低的汽气比，高的百分增量与高的汽气比有关，双工质循环的比功增加给人以深刻的印象。

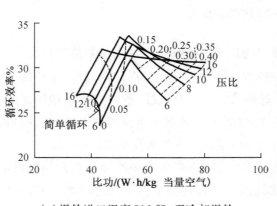

(a) 涡轮进口温度 816 ℃，无冷却涡轮

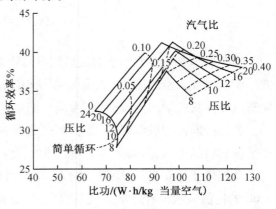

(b) 涡轮进口温度 1 093 ℃，气冷涡轮

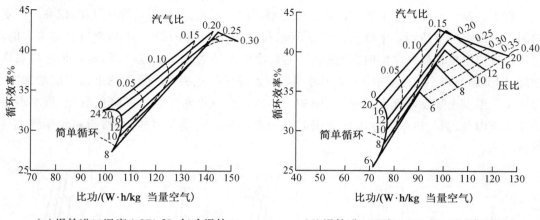

(c) 涡轮进口温度 1 371 ℃，气冷涡轮

(d) 涡轮进口温度 1 093 ℃，无冷却涡轮

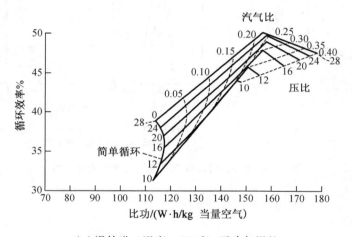

(e) 涡轮进口温度 1 371 ℃，无冷却涡轮

**图 3　双工质循环的循环效率和比功**

余热锅炉出口的排气温度与汽气比的关系示于图 4。在低的汽气比时，排气出口温度线很陡。高比值时，最小温差发生在极点位置，斜度就小多了。如果排气温度相当低，排气中水蒸气量较高，水蒸气将在锅炉出口附近凝结，图 4 中指出了发生这种现象的温度。从图中还可以看到这样一个现象：在同样的汽气比下，较低的压比导致一个较高的排气出口温度。

图 5 对一定的汽气比和压比给出了循环效率与涡轮进口温度的函数关系。图 5(a) 为无冷却涡轮，图 5(b) 为气冷涡轮。在较高的涡轮进口温度下，效率随压比增加而持续增加，涡轮进口温度较低时，较高压比的曲线与同样汽气比下较低压比的曲线相交。

由图 5 可见，对带无冷却涡轮的燃气轮机，效率随燃气初温增加而持续增加，对带气冷涡轮的燃气轮机，在等压比下有一相对较为平坦的峰值，这是由于随着燃气初温增加，所需的冷却量增加的结果。

双工质发动机在性能上表现出明显的差别。图5表明,对一固定的汽气比,仅在一特定的涡轮进口温度下达到效率峰值。从图5(a)可见,对无冷却的涡轮,其效率峰值是很尖的。当涡轮进口温度在峰值左侧时,余热锅炉的最小温差位置位于极点处,随着涡轮进口温度增加,端部温差减小,在达到峰值效率时端部温差等于极点温差。当涡轮进口温度增加时,由于排气温度增加及端部温差减小,锅炉出口的混合气能量增加。在峰值右侧,最小温差等于端部温差,涡轮进口温度增加会引起锅炉出口混合气温度增加,其值与涡轮排气温度增量相同。

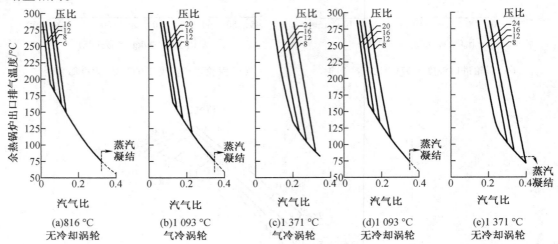

图4　余热锅炉出口排气温度与汽气比的关系图

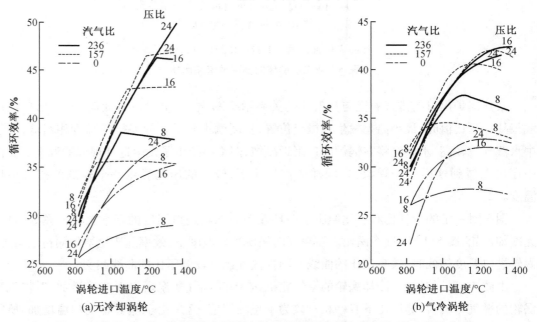

图5　涡轮进口温度对循环效率的影响(对于简单循环燃气轮机和双工质发动机)

对气冷涡轮并未显示在斜度上亦不连续。从图5(a)可见,对无冷却的涡轮,涡轮进口温度在峰值左侧时,效率的变化率几乎是常数。对气冷涡轮,随着涡轮进口温度增加,所需的冷却量也增加了,这使效率随涡轮进口温度的变化斜率减小。于是,气冷涡轮的循环效率 – 涡轮进口温度曲线成为光滑状曲线。

## 二、循环综合比较

文中选择了五种循环,就其性能、初投资、运行成本等方面加以比较。

A,简单循环燃气轮机。其相应于最大比功的压比为12,相应于最大效率的压比为30,在循环比较中取压比为16。循环效率33%。

B,燃–蒸联合循环。压比选为12,循环效率44%,比功为205 千瓦/磅[①]空气/秒。为保证达到最好的经济性,分析中的联合循环是由四台燃气轮机和一台蒸汽轮机组成。

C,喷注蒸汽的燃气轮机循环。考虑到目前单转子工业燃气轮机的水平,取压比为16。汽气比取为0.155,以满足排烟可见度方面的限制。

D,带回热的喷注蒸汽的燃气轮机循环。用燃气轮机的排气热量在压气机出口空气进入燃烧室之前进行回热。回热器之后有一余热蒸汽发生器,所产生的蒸汽喷入燃气轮机。当然,余热蒸汽发生器产生蒸汽的能力明显降低,结果其比功仅为C循环的90%。虽效率提高1%,但增加了回热器的成本,比功又下降,故此循环不作为一个经济的方案来考虑。

E,带背压式蒸汽轮机的喷注蒸汽的燃气轮机循环(图6)。余热蒸汽发生器中产生的高压蒸汽通过一背压式蒸汽轮机膨胀产生附加功率,然后喷入燃气轮机。与燃–蒸联合循环相比,E循环结构简单,无需冷凝器或冷却塔。比较中所选的压比为12,与压比16相比,其效率下降很小;与压比8相比,其比功下降很小。该循环的效率比C略好些,但比B差,其比功则是各比较循环中之佼佼者。

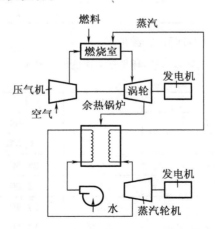

**图6　带背压式蒸汽轮机的喷注蒸汽的燃气轮机循环系统示意图**

---

① 　1磅(lb) = 0.453 592 4 千克(kg)(准确值)。

上述各循环的性能比较见表2。

表2　涡轮进口温度1 204 ℃的各循环性能

| 循环 | 压比 | 效率/% | 比功/(千瓦·(磅空气秒)⁻¹) | 汽气比 | 水耗/(磅水/(千瓦时)) | 蒸汽参数 | 限制条件 |
| --- | --- | --- | --- | --- | --- | --- | --- |
| A | 16 | 33 | 144 | 0 | 0 | — | 压比 |
| B | 12 | 44 | 205 | 0 | 1.69 | 1 450 psi/1 000 °F | 高效率 |
| C | 16 | 41 | 226 | 0.155 | 2.47 | 330 psi/952 °F | 排烟可见度 |
| D | 16 | 42 | 202 | 0.131 | 2.33 | 330 psi/834 °F | 排烟可见度 |
| E | 12 | 41 | 270 | 0.185 | 2.47 | 1 450 psi/1 000 °F | 余热锅炉和排烟可见度 |

各循环投资成本比较见表3。

表3　各循环的投资分配

（单位：百万美元）

| 循环 | A | C | E | B |
| --- | --- | --- | --- | --- |
| 燃气轮机－发电机 | 6.9 | 8.0 | 8.0 | 25.2 |
| 装置 | 1.9 | 1.9 | 1.9 | 7.6 |
| 余热蒸汽发生器和装置 | — | 4.1 | 5.0 | 20.0 |
| 蒸汽轮机—发电机和装置 | — | — | 3.1 | 16.0 |
| 水处理装置 | — | 0.3 | 1.1 | — |
| 冷凝器和冷却塔 | — | — | — | 5.5 |
| 不可预见费20% | 1.8 | 2.9 | 3.8 | 14.9 |
| 建造期利息 | 0.5 | 0.9 | 1.2 | 8.2 |
| 总投资 | 11.8 | 19.2 | 25.6 | 107.4 |
| 总投资/(美元/千瓦) | 138 | 142 | 158 | 202 |
| 建造期/年 | 1 | 1 | 1 | 2 |

各循环每千瓦时的电力成本比较见表4。

表4　各循环每千瓦时的电力成本

（单位：$10^{-3}$美元）

| 循环 | A | C | E | B |
| --- | --- | --- | --- | --- |
| 水（按30美分/1 000加仑） | — | 0.09 | 0.09 | 0.06 |
| 水处理 | — | 0.03 | 0.30 | — |
| 燃料（按3美元/百万英热单位） | 31.6 | 25.30 | 25.00 | 24.10 |
| 运行维护 | 2.0 | 2.00 | 2.00 | 2.00 |
| 可变成本（1974年） | 33.6 | 27.40 | 27.40 | 26.10 |
| 平均可变成本终值（电站寿命30年） | 65.2 | 53.10 | 53.20 | 50.70 |

表4(续)

| 循环 | | A | C | E | B |
|---|---|---|---|---|---|
| 总平均成本 | 年运行率 1.0 | 68.0 | 56.1 | 56.5 | 54.9 |
| | 年运行率 0.65 | 69.5 | 57.6 | 58.2 | 57.2 |
| | 年运行率 0.01 | 349 | 345 | 378 | 472 |

从比较中可见，E 循环和 C 循环的比功最大，各种喷注蒸汽循环的水耗大致在 2.5 lb/(kW·h)，比煤电站的 3.8 lb/(kW·h) 要小。C 循环的投资费用虽比简单循环燃气轮机略高，然而在电力成本方面具有明显的优点，在很低的年运行率下仍然如此。与燃-蒸联合循环相比，C 循环的投资费大大减少。虽然 B 循环的电力成本略微低一些，但到年运行率在 0.53 以下时 C 循环又表现出明显的优势。应强调指出，上述比较是在特定条件下进行的，但可看出，程氏循环在比功、效率、水耗、系统简易性、投资成本和电力成本等方面均有一定的优势，具有良好的综合指标。

## 三、应用实例

有关程氏循环发动机投入商用的报道，最早见于 1984 年。

国际动力技术公司的 Onan560-GTU 燃气涡轮发电机即是一例。该装置包括一个 Garrett IE 831-800 燃气涡轮发动机(压比 11，空气流量 7.9 lb/s，涡轮进口温度 1 760 ℉)和一个发电机，并配置了一个余热蒸汽发生器(包括过热器、蒸发器和经济器)。该余热蒸汽发生器能产生 3 000 lb/h 的最大蒸汽流量，其压力比压气机出口压力高 20 psi(在最大运行点)。城市供水通过去离子系统后泵入给水加热器。程氏循环发动机装置见图 7。

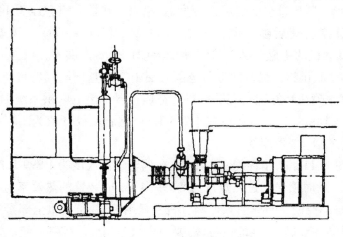

图 7　装置布置示意图

预计采用程氏循环后峰值功率将从 560 kW 增加到 1 100 kW，热效率为 34%。图 8 示出了系统效率与输出功间的试验关系。与通常的布拉东循环相比，程氏循环使在各负荷下的效率有大幅度增加,例如,输出功率为 400 kW 时,效率从原 17.5% 增至 23%,效率相对增加 31%。

图 9 示出了所测得的排温与输出功间的关系。在低负荷时程氏循环的排温要比布拉东循环低 50 °F,在输出功率为 400 kW 时,排温要低 100 °F。

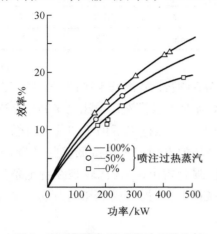

图 8　程氏循环与基础发动机的比较

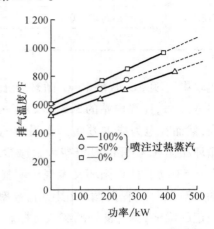

图 9　程氏循环与基础发动机的排温比较

试验表明,从慢车到原基础发动机的持续功率,程氏循环发动机相应地节约燃料 10% ~ 28%。

1984 年 11 月一台安装在美国加州大学的热电联供程氏循环发动机完成验收试验,并于 1985 年 1 月投入商业运行。其由 1 台 Allison 501 - KH 燃气轮机和余热锅炉组成。Allison 501 - KH 是由 Allison 501 - KB5 按喷注蒸汽运行的要求修改而成,其具有 6 个火焰筒的环管燃烧室,喷注用的过热蒸汽用两根直径 6 in① 的管子接到燃气轮机上。余热锅炉包括过热器、补燃室(紧接过热器之后)、蒸发器、给水加热器。汽包压力 14.4 kg/cm²,经过热器、管线和流量控制阀后喷射蒸汽压力为 11.9 ~ 12.6 kg/cm²。外输蒸汽的压力为 3.5 ~ 7 kg/cm²。发动机的主燃料是天然气,加压到 21 kg/cm² 供入。

图 10 表示热电联供程氏循环的运行范围。图中②为基础燃气轮机所产生的电功率和热输出(此时无蒸汽喷注和补燃)。从图可见,补燃扩展了热电联供系统的运行区。涡轮进口温度的设计值是 1 308 K,首台运行在 1 255 K,补燃温度设计值为 1 144 K,首台运行在 1 033 K,故而图上有实、虚两组线。

装置的布置见图 11。初步性能试验结果:在涡轮进口温度为 1 299 K 时,发电机的输出功率为 5 321 kW。此时喷注的过热蒸汽为 8 999.4 kg/h,装置的热效率为 37%。

整个电站全自动控制,包括运行、监测、保护、显示等,并可提供全部参数 26 h 的完整数据。为保证整个装置能以最佳性能运行,需要一个快速响应的控制系统。

---

① 1 in = 2.54 cm。

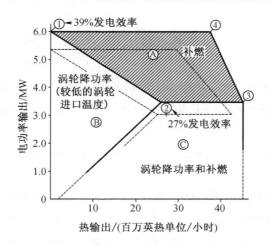

图 10　热电联供程氏循环运行范围

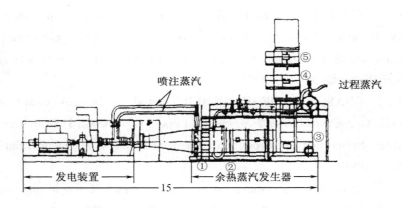

图 11　热电联供程氏发动机布置图

## 四、结束语

文中所综述的双工质平行－复合循环热机是一种基本系统,但并不意味着这是唯一可用的系统。依据使用场合和要求的不同,在此基本系统上加适当变化是完全可能的。

该循环除可在电站上应用外,其用途是广泛的,航空喷气发动机、机车、船舶、汽车发动机等都有应用的可能性。

需要消耗一定质量要求的处理水是这种循环最明显的缺点。然而,在水耗方面的比较,以及将水处理增加的成本与燃料费用及投资成本节省相比,至少在电站方面仍表现出其优势。

双工质平行－复合循环热机可在现有的燃气轮机上以较小的修改建成。在高度重视节能的当今,这又是一个重要的、有吸引力的经济因素。

## 参 考 文 献

[1] CHENG D Y. Parallel-compound dual-fluid heat engine:US 3978661[P]. 1976-09-07.
[2] CHENG D Y. Regenerative parallel compound dual-fluid heat engine: US 4128994[P]. 1978-12-12.
[3] CHENG D Y. Regenerative parallel compound dnal fluid heat engine: US 4248039[P]. 1981-02-03.
[4] WILKES C, RUSSELL R C. The effects of fuel bound nitrogen concentration and water iInjection on NOx emissions from a 75-MW gas turbine[C]//ASME 1978 International Gas Turbine Conference and Products Show. American Society of Mechanical Engineers Digital Collection, 1978.
[5] JONES J L, FLYNN B R, STROTHER J R. Operating flexibility and economic benefits of a dual-fluid cycle 501-KB gas turbine engine in cogeneration applications[C]//ASME 1982 International Gas Turbine Conference and Exhibit. American Society of Mechanical Engineers Digital Collection, 1982.
[6] JONES J L, CHANG C N, DIGUMARTHI R V, et al. Design and construction of the first commercial Cheng cycle series 7 cogeneration plant[C]//ASME 1985 Beijing International Gas Turbine Symposium and Exposition. American Society of Mechanical Engineers Digital Collection, 1985.
[7] Boyle R J. Effect of steam addition on cycle performance of simple and recuperated gas turbines[R]. Nosa Stilrecon Tecnical Report N, NASA, 1979:1-53.
[8] FARRELL R A, THOMAS M E. Effects of inlet conditions and water injection on emissions from a dual-fuel industrial gas turbine[C]//ASME 1981 International Gas Turbine Conference and Products Show. American Society of Mechanical Engineers Digital Collection, 1981.
[9] MULIK P R, SINGH P P, COHN A. Effect of water injection for NOx reduction with synthetic liquid fuels containing high fuel-bound nitrogen in a gas turbine combustor[C]//ASME 1981 International Gas Turbine Conference and Products Show. American Society of Mechanical Engineers Digital Collection, 1981.

# 燃气轮机 STIG 化的研究*

邹积国　闻雪友

（哈尔滨船舶锅炉涡轮机研究所）

**摘　要**：燃气轮机回注蒸汽(STIG)循环因其具有高效率、高比功及系统组成的简单性等特点受到广泛注意。能在现有的燃气轮机上实际 STIG 化是 STIG 发动机能迅速实用化的重要原因之一。文中讨论了对现有燃气轮机进行 STIG 化改装中的某些问题，并分析了在工业上采用单轴、双轴和三轴燃气轮机 STIG 化的工作情况，从中得出了一些结论。

## 一、引言

燃气轮机回注蒸汽(STIG)发动机在 20 世纪 80 年代中期已进入商用化阶段。目前投入商业运行的 STIG 机组都是在现有燃气轮机上改装而成的，一般用于热电联供。这种改装具有投资少、周期短和可靠性高的优点。因此，将现有的燃气轮机改成 STIG 发动机的研究具有很大的实际意义。

## 二、单轴燃气轮机的 STIG 化

单轴燃气轮机的 STIG 化首先要考虑以下两个因素：发动机允许的最大极限功率和自身的匹配。在发电装置中，单轴燃机以定转速运行，当考虑回注蒸汽增大功率时，发动机的强度问题应受到极大关注。另外，蒸汽的注入，尤其在汽气比较大时，破坏了压气机与涡轮的匹配，压气机压比势必增大，在定转速下压气机工作点移向喘振边界，同时涡轮的容积流量和膨胀比增加，涡轮工作点移向临界。

如果原机的涡轮处于亚临界状态，且压气机有较大的喘振裕度（工作在基本负荷下的燃气轮机、压气机一般都具有较大的喘振裕度），在原机极限功率允许的情况下，改成小汽气比 STIG 发动机，而不对原机通流部分进行调整是可能实现的。这种情况具有工程实际

---

\* 文章发表时间：1991 年 5 月。

意义。如在大气温度较高环境下工作的燃气轮机,在基本负荷下偏离设计工况较远,或者由于其他原因功率明显下降,改成 STIG 热电联供装置,不但可以使原机的功率恢复到 ISO 状态,热效率明显提高,而且回注蒸汽还可以作为外供热负荷的调节手段,使电厂运行的灵活性增强。整个系统构成简单、投资少、回收快,是很有吸引力的方案。

当原机涡轮设计工况处于临界状态时,除非允许压气机压比有较大幅度的增加,否则实施 STIG 化将十分不利。首先讨论压气机增压比不变的情况。由于涡轮临界,其折合流量已达最大,蒸汽的注入必使其涡轮入口温度($TIT$)大幅度下降,其结果随着蒸汽注入量的加大,性能越来越差(图 1)。如果允许压气机增压比在一定范围内增加,此时压气机的空气流量减小,压比升高,使得在一定涡轮入口温度下可以加入相当量的蒸汽而获得较高的性能。但是,此时应该注意处于临界状态下的涡轮是否有能力完成增大后的膨胀比。

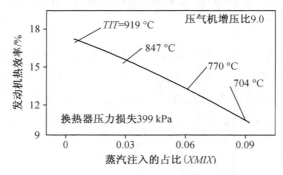

图 1  蒸气注入对发动机热效率的影响

对于单轴燃气轮机,如果不对原发动机进行调整,要达到最佳汽气比下的回注蒸汽量十分困难。由于发动机极限功率的限制、压气机喘振裕度和涡轮做功能力的限制,要综合解决这些矛盾,意味着发动机有相当的改动量。

首先讨论一台小型单轴燃气轮机。在大气温度 15 ℃、大气压力 0.101 3 MPa、无排气损失时,发动机的性能参数如下:涡轮入口温度为 930 ℃,压气机增压比为 9,压气机入口空气流量为 1.75 kg/s,发动机输出功率为 223 kW,发动机排气温度为 520.5 ℃,发动机热效率为 18.3%,发动机按定转速运行。

图 2 给出了 STIG 发动机设计点选择的总性能曲线(涡轮定几何)。可见,如不考虑发动机本身的限制条件(极限功率、喘振裕度、涡轮工作状态),该机 STIG 化后确实具有较高的性能。在涡轮入口温度为 930 ℃时,最佳热效率达 32.3%。但是,由于限制条件存在,发动机仅能在低于最佳效率点以下区域运行。如果发动机的极限功率较大,则 STIG 发动机性能亦较高,在允许的压气机喘振裕度下,涡轮入口温度为 930 ℃时,其热效率为 24.9%。但是,随着汽气比的增加,压气机的喘振裕度限制成了主要矛盾,此时可以通过调整涡轮导向器面积来使压气机的喘振裕度在合理的范围内。

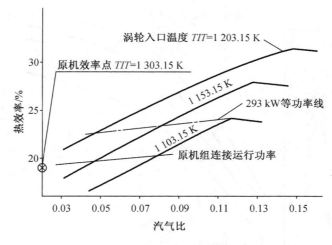

**图 2　STIG 发动机设计点选择的总性能曲线**

## 三、双轴燃气轮机的 STIG 化

双轴燃气轮机在大气温度 28 ℃、大气压力 0.101 3 MPa 时,其性能参数如下:涡轮入口温度为 815 ℃,压气机增压比为 6.30,压气机入口空气流量为 36.8 kg/s,发动机输出功率为 4 400 kW,发动机热效率为 18.5%,燃气轮机为分轴式。

在发动机的涡轮入口温度、压气机增压比重部件效率不变,且其 STIG 循环的最佳效率不变的情况下,其 STIG 循环的最佳效率为 30.9%,而相应于最佳压比下的最佳热效率为 31.4%。由此可见,由发动机压气机的增压比偏离 STIG 最佳压气机压比所造成的热效率下降仅为 1.6%。因此,用这台机组进行 STIG 化可望达到较好的性能。

图 3 给出了在最佳效率下压气机的喘振裕度随高压涡轮第 I 级导向器转角的变化曲线。当涡轮不变几何时,在相应于最佳效率点的汽气比下,压气机的喘振裕度都很小,发动机不能正常运行。因此,要综合权衡下列诸因素后才能确定涡轮的几何条件:①压气机的喘振裕度;②与动力涡轮的匹配;③热效率和输出功率。分析结果表明,在高压涡轮改动量不大(例如 I 级导向器转角 10°)的情况下,能源满足燃气发生器的匹配,但动力涡轮的矛盾突出。由于蒸汽的注入,动力涡轮的膨胀比大幅度增加,而入口折合流量减小,要满足动力涡轮与燃气发生器的匹配,动力涡轮的改动量相当大,甚至要重新设计。

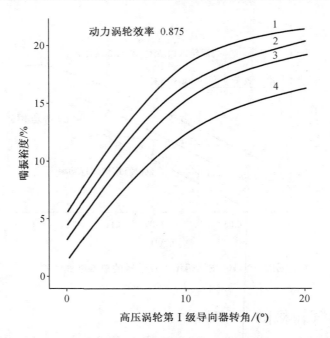

**图 3** 最大效率下压气机喘振裕度随高压涡轮第 I 级导向器转角的变化曲线(相对转速 1.0)

1—涡轮入口温度为 1 033.15;2—涡轮入口温度为 1 053.15
3—涡轮入口温度为 1 073.15;4—涡轮入口温度为 1 093.15

图 4 给出了燃气轮机 STIG 化后的总性能曲线。当涡轮入口温度为 800 ℃时,发动机输出功率为 6 960 kW,热效率为 26.5%。与原机相比,热效率提高了 43%,功率提高了 56%,并且涡轮入口温度下降了 15 ℃,有助于提高发动机的寿命。

文中所讨论的两台单、双轴燃气轮机均属于第一代燃气轮机,其部件效率较低,性能差。显而易见,性能较差的燃气轮机在改成 STIG 发动机后,其相对收益就越加明显。

另一个较为重要的问题是,涡轮工质变化对涡轮性能的影响在 STIG 燃气轮机中分析部件匹配时是很重要的。图 5 和图 6 给出了在同一涡轮入口条件下不同工质(燃气、蒸汽与燃气的混合气体)的涡轮特性曲线。在同样的膨胀比 $\delta_T$ 和相对转速 $\lambda_\mu$ 下,工质变化后,相应的涡轮折合流量减小。如果要使折合流量保持不变,必须使涡轮的膨胀比增加。这一点在进行双工质发动机的设计过程中很重要。

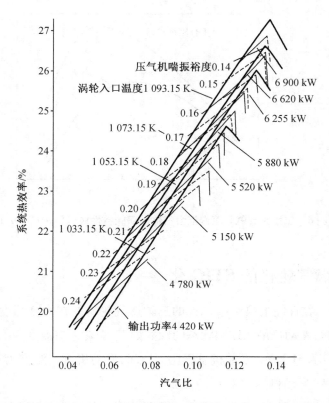

图 4 双轴 STIG 发动机燃气发生器性能曲线(高压涡轮等 I 级导向器转角 10°(开关),燃气发生器相对转速 1.0)

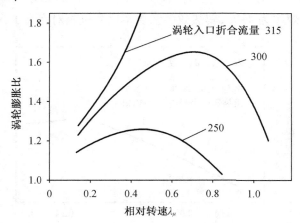

图 5 工质为燃气的涡轮特性曲线

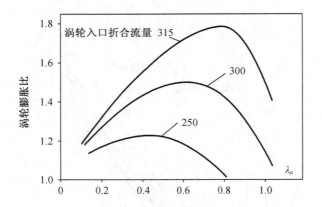

图6　工质为燃气与蒸汽混合气体的涡轮特性曲线（$XMIX = 0.128$）

## 四、三轴燃气轮机的 STIG 化

现在讨论一台在性能上属于第二代的三轴燃气轮机。在大气温度为 27 ℃，大气压力为 0.101 3 MPa 时，发动机最大连续状态的性能如下：涡轮入口温度为 1 052 ℃，低压压气机入口空气流量为 47.51 kg/s，发动机总压比为 15.6，发动机排气温度为 481 ℃，发动机输出功率为 10 000 kW，发动机热效率为 29.8%。

图7给出了 STIG 发动机设计点选择中热效率随汽气比变化的曲线，最佳汽气比下对应的参数为：涡轮入口温度为 1 052 ℃，最佳效率下的汽气比为 0.123，发动机输出功率为 16 000 kW，发动机热效率为 41%，其热效率和功率比原机相对提高了 38% 和 61%。

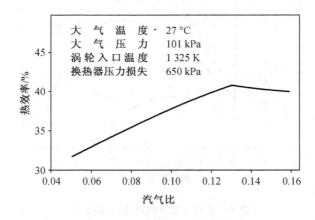

图7　热效率随汽气比变化曲线

三轴燃气轮机 STIG 化中的重要问题与双轴类似。燃气发生器改动量较小，通过高低压涡轮导向器面积的调整来实现 STIG 发动机的燃气发生器匹配（本方案中，仅低压涡轮第

Ⅰ级导向器开大了 3.9°），动力涡轮则不能正常工作。蒸汽注入使压气机涡轮膨胀比减小，而动力涡轮的膨胀比增加了 25%，折合流量则减少了 13%，因此，动力涡轮需要进行较大改动。由此看来，以动力涡轮输出功的燃气轮机，在改为 STIG 发动机时，动力涡轮与燃气发生器的匹配是主要问题。

在双轴或三轴 STIG 发动机中，如燃气发生器转速允许调整提高，那么对 STIG 化最为有利。选择 STIG 设计点时，可使在压气机工作点沿共同工作线向上移动，压气机涡轮略做调整就可能使燃气发生器满足匹配。

为了减小 STIG 化中燃气发生器与动力涡轮的调整工作量，可以考虑采用"多点"加注蒸汽方案。分别从高压涡轮、低压涡轮（也可作为冷却介质）和动力涡轮前注入蒸汽，其结果缓解或解决了燃气发生器与动力涡轮的匹配矛盾以及较大的蒸汽注入量问题，但性能收益有所下降。对于双轴 STIG 发动机，从动力涡轮前注入蒸汽，可以从根本上解决匹配问题。若动力涡轮的强度允许，且压气机具有足够的喘振裕度，通过改变燃室和动力涡轮前注入蒸汽的比例，不经改动就可能完成 STIG 化的实施，但收益约下降 50%。三轴燃气轮机采用"多点"加注蒸汽方案时，燃气发生器的调整量将有所增加。图 8 给出了 STIG 发动机的热效率在总注汽量不变的条件下随着非燃烧室部位注汽量增加而变化的情况。可见，效率随之下降（相对于全部蒸汽在燃烧室注入而言）。在动力涡轮前注入，效率最低，但对缓解匹配效果最佳。因此，"多点"注入蒸汽方案，尽管效率增益相对较低，但通过合理的配置有可能使涡轮部分的改动量趋于最小。

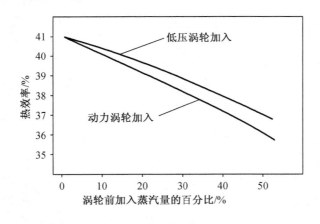

图 8 "多点"加入蒸汽量的性能比较

## 五、结论

1. 单轴燃气轮机改为 STIG 发动机中的主要问题是压气机喘振裕度和极限功率的制

约,要获得大回注量、高性能的STIG发动机,单轴燃气轮机的改动量将较大。

2. 如果单轴燃气轮机的涡轮处于临界状态,除非压气机具有较大的喘振裕度,涡轮临界没有达到极限,否则不宜改为STIG发动机。

3. 双轴或三轴燃气轮机改为STIG发动机的工作中最为突出的问题是动力涡轮与燃气发生器的匹配问题。当追求最佳性能时,动力涡轮的改动量大。

4. 在燃气轮机的STIG化中,采用"多点"注入蒸汽方案可以缓解在第3点中所述的动力涡轮改动量大的矛盾,但性能收益下降。

5. 涡轮工质变化对涡轮特性的影响不可忽视。

## 参 考 文 献

[1] CHENG D Y. Regenerative parallel compound dual – fluid heat engine:US 4128994[P]. 1978-12-12.

# 燃气轮机回注蒸汽装置的研究*

闻雪友　金介荣　傅　正　邹积国
（哈尔滨船舶锅炉涡轮机研究所）

**摘　要**：介绍了我国第一套燃气轮机回注蒸汽整机实验装置及其部件、设计的主要特点以及初步性能试验结果。

## 一、引言

自国际上第一套双工质平行－复合循环（即回注蒸汽燃气轮机循环 STIG）商用装置于 1985 年 1 月正式投入商业运行以来，短短数年间，这种 STIG 装置获得了迅速的发展。目前在国外主要用于热电联供装置和发电装置。其具有的高比功、高效率以及在满足热、电负荷平衡方面所具有的高度灵活性和投资费用低的特点使其受到青睐。

美国海军在讨论未来的护卫舰、驱逐舰和巡洋舰的驱动燃气轮机时认为有两种技术途径将有利于未来航机舰改的商品化，并且在技术上是可行的。其中之一就是直接在燃气轮机内利用余热锅炉产生的蒸汽，即双工质平行－复合循环，以简化 COGAS 的 RAC－ER 设计概念。虽然其对水质的要求高，但对于这一基本概念的某些方案来说，还不是很大的限制因素[1]。

哈尔滨船舶锅炉轮机研究所和哈船院联合，利用一台工业小燃气轮机，设计配置了蒸汽发生器、蒸汽回注系统、控制系统等，实现了燃气轮机回注蒸汽装置的整机实际运行。

## 二、STIG 装置

整个试验装置如图 1 所示。一台 SIA－02 工业燃气轮机与一台蒸汽发生器相组合，将蒸汽发生器产生的蒸汽回注入燃气轮机的燃烧室中，使基础发动机的出力大大增加，循环效率也随之增高。

---

\* 文章发表时间：1991 年 5 月。

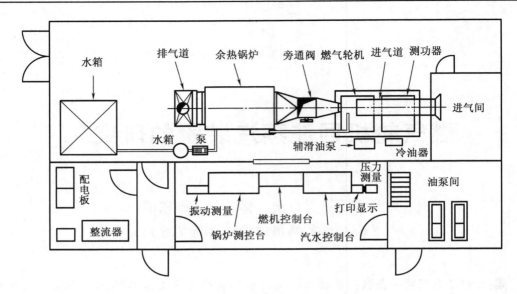

**图 1 试验装置图**

（一）燃气轮机

基础发动机是日本川崎重工株式会社生产的 SIA-02 发动机,是一台持续功率为 221 kW,备用功率为 228 kW 的工业小燃气轮机,由两级离心式压气机和两级轴流式涡轮以及单管回流式燃烧室组成。设计点的压气机压比为 9,空气流量为 1.8 kg/s,燃气初温为 930 ℃,涡轮转速为 53 000 r/min。通过一个两级减速的减速器,由前端输出功率,输出轴转速 1 500 r/min。发动机调节方式为等转速调节。发动机外形尺寸为 1 087 mm × 1 067 mm × 843 mm,燃油为柴油,滑油为合成润滑油。在 STIG 装置上,对发动机的燃烧室外壳做了修改,在与火焰筒主渗混孔相对应的部位设计了三个回注蒸汽接口,并相应地装有蒸汽喷嘴,蒸汽和燃气在渗混区混合后进入涡轮,膨胀做功。

（二）蒸汽发生器

蒸汽发生器利用燃气轮机排气余热产生过热蒸汽,作为燃气轮机另一种工质。

蒸汽发生器为直流式锅炉,无汽包。由预热段、蒸汽段、过渡段和过热段四部分组成。过热段直接位于燃气轮机排气扩压段之后,在回注前提高蒸汽温度,可减少为使蒸汽温度提高到燃气初温所需输入的燃料量,从而提高发动机的热效率。过渡段、蒸发段和预热段依次排列于后。

整个蒸汽发生器采用卧式布置形式。

（三）烟道

在燃气轮机与蒸汽发生器之间设置旁通烟道,利用排气换向阀可使燃气轮机的排气不经蒸汽发生器而直接从旁通烟道排出,此时燃气轮机的背压亦较低。在蒸汽发生器出口处有主排气烟囱。

## (四)水、汽系统

锅炉用水为除盐水。考虑到蒸汽发生器在投入运行前需洗炉,在正常维修保养时需进行冲洗,因此在水系统中设置必要的阀门和管道,以便进行酸洗、碱洗、正冲洗和反冲洗。

图2为蒸汽系统示意图。注入燃气轮机的蒸汽通过主蒸汽管道、压力脉冲阻尼器、蒸汽流量孔板、调节阀、速关阀、单向阀、蒸汽分配器,最后经过三根金属软管到三个带滤网的蒸汽喷嘴组件而进入燃烧室。为防止凝结水进入燃烧室,在管路上设有加温、吹扫通道,并设置了吹扫阀。蒸汽旁通管路系统由闸阀、蒸汽流量孔板、调节阀等组成。试验装置运行时,旁通系统将蒸汽发生器产生的过量蒸汽排到大气中。

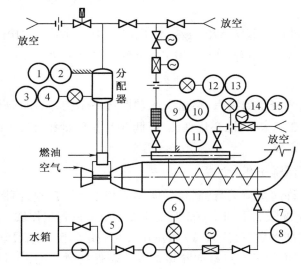

**图 2　蒸汽系统图**

$1-\frac{T_3}{Z}$；$2-\frac{T_3}{JZ}$；$3-\frac{P_3}{JZ}$；$4-\frac{P_3}{ZZ}$；$5-\frac{P_0}{Z}$；$6-\frac{G_1}{Z}$；$7-\frac{P_1}{Z}$；$8-\frac{T_1}{Z}$；

$9-\frac{T_2}{Z}$；$10-\frac{T_2}{JZ}$；$11-\frac{P_2}{Z}$；$12-\frac{G_3}{Z}$；$13-\frac{G_3}{JZ}$；$14-\frac{G_3}{Z}$；$15-\frac{G_2}{ZJ}$

## (五)调控监测系统

本试验装置的调控监测系统由两部分组成。

**1. 燃气轮机调控监测系统**

燃气轮机的燃料控制系统为电子模拟式燃料控制系统,主要完成启动加速过程、调速、非正常状态的报警停机控制,以保证燃气轮机在稳定和过渡态等各种运行状态下安全可靠地运行,并设有监测仪表板。

**2. 第二工质——水蒸气调控测量系统**

该系统以 DDZ - Ⅲ型仪表和必要的接口、转换电路,DJS - 033 微型计算机完成蒸汽发生器和蒸汽注入系统的操作控制、热工参数的检测、显示(包括系统模拟图)记录、整理和打印等功能。

### （六）测功系统

功率吸收和测量由 D650 型水力测功器完成。

## 三、设计特点

### （一）采用直流余热锅炉

此种锅炉与常规的汽鼓锅炉相比，金属耗量小、质量轻、体积小、造价低。直流锅炉的接头少，体积小，造价亦低。直流锅炉的接头少，简单、容水量小，而且水和蒸汽的全部容量也较小，增加了安全性。该锅炉对负荷变化的响应也较迅速；缺点是其对给水的水质要求较高。

锅炉在设计点下的产汽量为 570 kg/h，额定压力为 $1.079^{+0.15}$ MPa，过热蒸汽温度 413 ℃，锅炉烟气阻力 1.5 kPa。在结构上采用了螺旋鳍片管。

### （二）注汽方式

针对单管回流式燃烧室的特点，尽量少改动燃烧室结构，采用从燃烧室掺混孔正交喷注的方案。这是一个在熄火特性、燃烧性能，降低 $NO_x$ 排放量和废气中的 CO 及 HC 含量等各因素间折中的一个方案。

### （三）实验装置的试验基准设计点

由于发动机不允许超负荷运行，涡轮导向器面积不允许调整，而涡轮又基本上是在临界状态下工作，这些情况限制了 SIA-02 双工质发动机的可行工作域。因而，要获得高性能是一件困难的事。在高参数下，注入蒸汽的同时，为满足匹配要求，压气机压比势必增大。在定转速运行下，压气机的工作点移向喘振边界。若保持燃气初温不变，则发动机功率将大大超过原机极限功率。所以，在一定的涡轮工作状态下，应遵守以下三个原则：

（1）保证压气机有适宜的喘振裕度。

（2）注蒸汽后的最大功率为燃气轮机原机的全功率。

（3）燃气初温虽必须下降，但不宜太低，这确定了 SIA-02 双工质平行-复合循环装置试验的基准设计点：压比 9.2；燃气初温 1 087 K；汽气比为 0.072。

### （四）注汽系统的防水

由蒸汽发生器产生的过热蒸汽注入燃烧室的整个系统中，要防止可能有水进入工作中的燃气轮机，在向燃烧室回注蒸汽之前，管路也必须是热的。单向阀的管路用过热蒸汽来加温，并通过吹扫阀吹除杂质和冷凝水。单向阀后的一段用燃气轮机压气机出口的热空气来加温，热空气通过节流小孔排放至大气，蒸汽注入后该孔由电磁阀关闭。整个注汽系统用绝热材料包覆。

## 四、性能试验结果

双工质平行-复合循环系统是一个由燃气轮机和蒸汽发生器组成的动力耦合系统，回注过热蒸汽的反馈性质是燃气轮机性能与蒸汽发生器性能间的关键联系。正确的关联可使系统中的各部件在其各自的良好状态下运行。

对发动机功率1/8~8/8,汽气比2%~8%进行了大量的试验,得到了一系列的试验结果,建立了注蒸汽量、燃气温度、功率热效率等参数间的关联。

图3典型地示出了主要试验结果。在一定的燃气初温下,随着注汽量的增加,功率迅速增加,热效率也随之提高。热效率随汽气比增加而增加的情况一直持续到达到最佳汽气比,此后汽气比继续增长,循环效率将下降。另一个特点是随着燃气初温的增高,性能更好,这主要是由燃气轮机的基本特性所决定的。

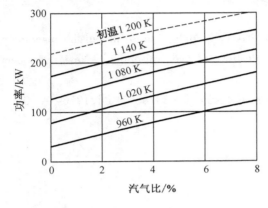 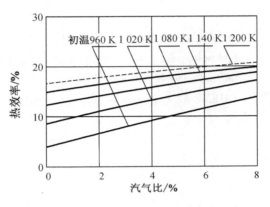

(a) SIA-02 STIG 的功率与汽气比的关系　　　(b) SIA-02 STIG 的热效率与汽气比的关系

图3　试验结果

从试验结果可明显看出,双工质平行-复合循环在增加功率、改进循环效率方面的效果是极其明显的。

图4给出了设计注汽量(7.2%)下,理论计算结果与试验结果的对比,两者在总体上非常吻合,得到了预想的结果。

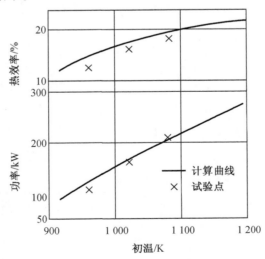

图4　SIA-02STIG 变工况性能(汽气比=0.072)

## 五、结束语

双工质平行-复合循环装置的实际运行试验表明:燃气轮机回注蒸汽实现了预期的功率和循环效率的大幅度提高。回注蒸汽已成为该领域中的一项重要新技术,这一新概念在我国亦已具备了进入工程实践的条件。

## 参 考 文 献

[1] BASKERVILLE J E, QUANDT JR E R, DONOVAN M R. Future propulsion machinery technology for gas turbine powered frigates, destroyers, and cruisers[J]. Naval engineers journal, 1984, 96(2): 34-46.

[2] CHENG D Y. Pavallel-compound dual-fluid heat engine: US 3978661[P]. 1976-09-07.

# 燃气轮机湿空气回注循环分析[*]

闻雪友　陆　犇　李名家

(哈尔滨船舶锅炉涡轮机研究所)

**摘　要**：讨论燃气轮机湿空气回注循环，提出将循环分为内部和外部空气加湿回注循环两类。以部分空气回热回注循环(PRSTIG)为基础，分析了燃气初温、压比、回注比、回热比等参数对循环效率、比功的影响。通过对两类各相关循环的特点的比较、讨论，得出结论：湿空气回注循环可使功率提高 10% ~ 25%，热耗降低 6% ~ 15%，$NO_x$ 排放量降低 15% ~ 50%，而且均可在现行装置上改造实施。

## 一、引言

众所周知，回热燃气轮机循环具有高效率，而蒸汽回注循环具有高比功，将回热燃气轮机循环与燃气轮机回注蒸汽(STIG)循环结合起来，形成一种新的复合循环——湿空气回注循环。其又可分为两类，即燃气轮机压气机出口空气(或压气机出口部分空气)被加湿、回热、回注的燃气轮机循环(内部空气加湿回注循环)和对由外部提供的压缩空气进行加湿、回热、回注的燃气轮机循环(外部空气加湿回注循环)。上述两类循环均有提高功率、降低热耗、减少 $NO_x$ 排放的优点。

## 二、内部空气加湿回注循环

### (一)循环原理

典型的循环系统如图 1 所示。工质空气经压气机压缩后分为两部分：一部分空气直接进入燃烧室，另一部分空气进入喷射器，经蒸汽引射后与之混合，形成的混合气体在余热锅炉内被燃气轮机的排气加热(回热)，然后注入燃烧室。另一种工质水进入余热锅炉，从排气废热中吸收热量，水被加热成饱和蒸汽，然后进入喷射器与空气混合。从燃烧室出口的燃气进入涡轮做功后，涡轮排气在余热锅炉内首先加热湿空气，然后用于产生饱和蒸汽。

---

[*] 文章发表时间：2006 年 1 月。

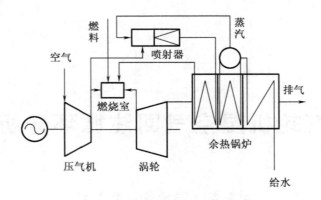

**图 1　PRSTIG 循环系统示意图**

该循环的特征是部分空气回热注蒸汽,简称 PRSTIG(partial regenerative steam injected gas turbine)循环。

实际上,当压气机出口空气全部先进入混合器加湿后再进入燃烧室时,则形成回热注蒸汽燃气轮机循环,简称 RSTIG 循环。显然,RSTIG 循环可以看作 PRSTIG 循环分析中的一个特例。反之,如果压气机出口的空气全部直接进入燃烧室时,则形成 STIG 循环。

循环间的关联形式如图 2 所示。

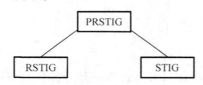

**图 2　循环间的关联形式**

## (二)循环分析

为了更具体地说明 PRSTIG 循环,文中进行了计算分析,分析中所用的主要参数值(假定值)列于表 1。文中研究了某些参数在一定范围内变化的影响,此时其他参数值一般保持常数。

**表 1　PRSTIG 循环系统参数表**

| 参　　数 | 数　　值 |
| --- | --- |
| 环境温度/℃ | 20 |
| 环境压力/$10^5$ Pa | 1.013 25 |
| 给水温度/℃ | 20 |
| 燃料注入温度/℃ | 25 |
| 压气机效率 | 0.88 |
| 涡轮效率 | 0.90 |

表1(续)

| 参　数 | 数　值 |
|---|---|
| 燃烧效率 | 0.99 |
| 机械效率 | 0.985 |
| 排气温度最低值/℃ | 200 |
| 回热器回热度 | 0.85 |
| 余热锅炉节点温差/℃ | 15 |

图3表示 PRSTIG 循环热效率与回热比 $R_a$(回热空气质量流量与压气机进口空气质量流量的比值)、回注比 $R_s$(回注蒸汽质量流量与压气机进口空气质量流量的比值)之间的关系。回注比最佳值将曲线分为两支,上升曲线表示此时涡轮排气中有足够热量将工质加热到所要求的参数,这时空气与蒸汽组成的混合气体有较好的过热度;下降曲线表示此时因回注比偏高,只好以降低注气温度(蒸汽与空气所组成的混合气体注入燃烧室时的温度)作为补偿。这时混合气体在燃烧室内所需的吸热量增加,导致效率下降。

图4表示 PRSTIG 循环比功与回注比、回热比的关系。当回热比保持一定时,回注比的增加使进入涡轮的工质质量流量增加,同时单位质量的工质做功能力也增加,故比功也随之增加。当回注比保持不变时,随回热比增加,比功略微有所下降。

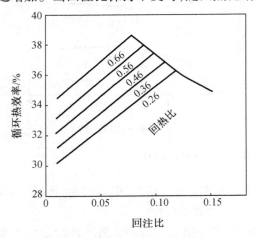

图3　$R_a,R_s$ 对循环热效率的影响(燃气初温 1 273.15 K,压比8)

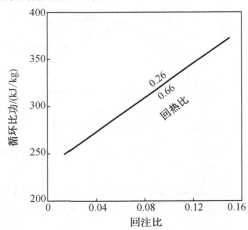

图4　$R_a,R_s$ 对循环比功的影响(燃气初温 1 273.15 K,压比8)

图5表示 PRSTIG 循环最大热效率(此时回注比处于最佳值)与压比、回热比之间的关系。随着压比的升高,回热温差减小,回热比的变化对循环最大热效率的影响程度减弱。因为压比适当下降可使回热温差升高,对回热有利,故回热比越大,最佳压比越小。

图6表示 PRSTIG 循环最佳回注比(此时循环具有最大热效率)与压比、回热比之间的关系。

图7表示循环性能与燃气初温之间的关系。由图可见,燃气初温仍是影响 PRSTIG 循环性能的主要因素。

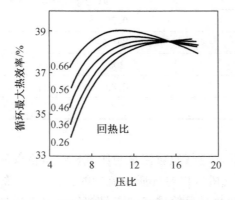

**图5** $R_a$、压比对循环最大热效率的影响(燃气初温 1 273.15 K)

**图6** $R_a$、压比对循环最佳回注比的影响(燃气初温 1 273.15 K,压比 8)

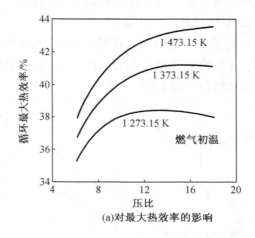
(a)对最大热效率的影响

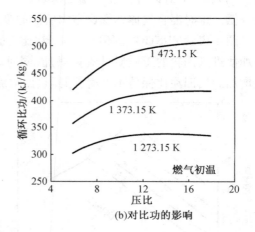
(b)对比功的影响

**图7** 燃气初温对循环性能的影响(回热比 0.46)

综上所述,影响和决定 PRSTIG 循环性能的独立参数主要有四个:燃气初温、压比、回注比及回热比。PRSTIG 循环性能的获得,取决于这四个参数的匹配。

为了更深入地了解循环的特点,将 STIG 循环、PSTIG 循环、PRSTIG 循环性能加以简单对比,计算参数选自表1,同时取燃气初温 1 273.15 K,回热比 0.46。

图8表示 STIG 循环、RSTIG 循环、PRSTIG 循环最大热效率与压比之间的关系。由图中可以看出:每种循环均存在一最佳压比,此时循环具有最大热效率的最佳值,并且最佳压比的关系为 STIG > PRSTIG > RSTIG。由于回热的影响,当压比超过某一值 $\pi_1$ 后,RSTIG 循环的最大热效率低于 PRSTIG 循环的最大热效率;压比继续升高到另一值 $\pi_2$ 后,PRSTIG 循环的最大热效率低于 STIG 循环的最大热效率。

图9表示三种循环的最大热效率与最佳回注比间的关系(此时压比固定为8)。

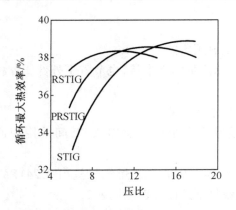 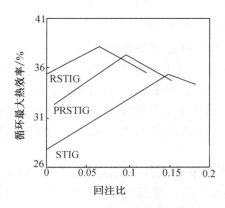

图 8 压比对循环最大热效率的影响　　　图 9 回注比对循环热效率的影响

## 三、外部空气加湿回注循环

### （一）循环原理

典型的循环系统如图 10 所示。空气由外部的、单独的用电机驱动的压缩机（带间冷器）增压进入空气饱和器，空气被湿化后经热回收装置，利用燃气轮机排气废热进行预热后注入燃气轮机燃烧室的上游。另一种工质（水）进入热回收装置，从排气废热中吸收热量，水被加热成饱和蒸汽，然后进入空气饱和器对增压空气加湿。从燃烧室出口的燃气进入涡轮做功后，涡轮排气在热回收装置内首先加热湿空气，然后用于产生蒸汽。

该循环的特征是外部空气加湿预热，故简称 HAI(humid air injection) 循环。

实际上，当外部增压空气不经加湿而仅在热回收装置中预热后即注入燃烧室时，则形成干空气回注循环，简称 DAI(dry air injection) 循环。

显然，DAI 循环可以看作 HAI 循环分析中的一个特例。

此外，如果外部增压空气为零，则形成 STIG 循环。

循环间的关系形式如图 11 所示。

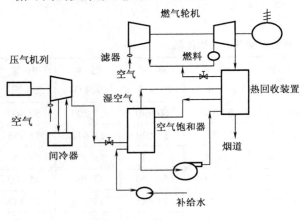 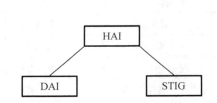

图 10　HAI 系统示意图　　　　　　　图 11　循环间的关系

## (二)循环分析

(1) HAI 系统功率的增量与注入的湿空气流量成正比。

(2) 相对于干空气回液压,注湿空气可降低对总流量的需求量,也节省了外部压缩机的功耗(否则需对更大的空气流量增压)。

(3) 注湿空气的效果(功率、热耗率)明显优于注干空气。

(4) 为尽可能降低外部空气增压的功耗,通常用多个压气机串联,其间采用中间冷却器。

为更具体地说明,引用美国南加利福尼亚大河能源中心 3 号燃气轮机(PG7241FA)进行说明,其在环境温度 35 ℃,海平面及不同加湿条件下的设计性能见表2。

表 2  不同加湿条件下 PG7241FA 性能

| 性能 | 湿空气 | 湿空气 | 干空气 | 现装置(PG17241FA) |
|---|---|---|---|---|
| 加湿度/% | 7.5 | 5.5 | — | — |
| 净功率/MW | 182.6 | 177.0 | 166.4 | 150.4 |
| 热耗率/kJ | 9 664 | 9 896 | 10 255 | 10 297 |

该机已进行了 HAI 循环的初期验证试验,试验是在 35 ℃,3.5% 的加湿量下进行的,这是 GE 公司对 Fr7FA 系列燃气轮机蒸汽回注增功的限定,涡轮转子入口温度也有所降低(按"干"控制线运行),试验结果表明功率增加 18.3 MW。当按"湿"控制线运行时,功率可增加近 22.3 MW。

为更进一步说明,列举了相关数据(表3),数据取自 Hill 能源系统和 PB 动力提出的一份在 4 台 GE Fr7B 调峰机组上装湿空气回注装置的建议书。表中表示了现运行的 Fr7B 燃气轮机,其净功率按平均降低 8% 计(由于长期运行),以天然气为燃料,环境温度 29.4 ℃ (85 ℉)为设计点,无湿空气回注与 7.6% 湿空气回注情况下性能的全面对比。从增益栏中可以看出,所获的增功部分是在极高的效率下实现的。

表 3  Fr7B 注湿空气的性能预测

| 性能 | 25 ℉ (-3.9 ℃) | 40 ℉ (4.4 ℃) | 59 ℉ (15 ℃) | 80 ℉ (26.7 ℃) | 85 ℉ (29.4 ℃) | 95 ℉ (35 ℃) |
|---|---|---|---|---|---|---|
| HAI 电站净出力/kW | 72 820 | 69 500 | 65 540 | 60 940 | 59 800 | 57 400 |
| 热耗率/kJ | 11 014 | 11 130 | 11 310 | 11 579 | 11 647 | 11 795 |
| 效率/% | 32.7 | 32.4 | 31.8 | 31.1 | 30.9 | 30.5 |
| 现电站净出力/kW | 59 200 | 56 200 | 52 300 | 48 200 | 47 300 | 45 400 |
| 热耗率/kJ | 12 702 | 12 855 | 13 093 | 13 483 | 13 573 | 13 747 |
| 效率/% | 28.3 | 28.0 | 27.5 | 26.7 | 26.5 | 26.2 |

表3(续)

| 性 能 | 25 ℉<br>(-3.9 ℃) | 40 ℉<br>(4.4 ℃) | 59 ℉<br>(15 ℃) | 80 ℉<br>(26.7 ℃) | 85 ℉<br>(29.4 ℃) | 95 ℉<br>(35 ℃) |
|---|---|---|---|---|---|---|
| 增益净出力/kW | 13 600 | 13 300 | 13 200 | 12 800 | 12 500 | 12 000 |
| 热耗率/kJ | 3 661 | 3 809 | 4 262 | 4 415 | 4 410 | 4 405 |
| 效率/% | 98.4 | 94.5 | 84.5 | 81.5 | 81.6 | 81.7 |

## 四、现实可行性

文中讨论的循环特点是均可在现有装置上改造实施。在设计和机械方面有些类似STIG装置,也有功率增量不能超过运行极限的制约,最大允许扭矩的制约,以及压气喘振裕度、燃气温度、火焰稳定性和发电机容量的制约。

对于没有装干式低$NO_x$燃烧室的燃气轮机,注湿空气的效果更好,因为非干式低氮燃烧室允许在更高的含水量下运行,这导致更高的相对功率增量和更低的热耗。不仅如此,注湿空气也成为一种低成本的控制$NO_x$的技术。在Fr7B燃烧室上的试验表明,燃用天然气并注湿空气后,$NO_x$排放从100 mg/kg降至50 mg/kg以下。

注干空气不如注湿空气有效,但注干空气对部件寿命完全没有影响(与蒸汽回注相比)。注空气的功率增量相对较小,但在缺水的地方(如天然气管线增压站)可能是有吸引力的。因为这是一个扩充增压站容量而投资较少的方法,不必另外增加燃气轮机。对有多台燃气轮机装置的增压站,干空气回注系统可最佳化服务于总的增功需求,可明显降低成本,增加系统的可靠性、可用性和灵活性。

实际上文中所讨论的循环介于STIG与HAI之间,各有各的特点及各自的适用场合,PRSTIG在日本已有正式产品及应用,HAI在美国已进行了实机试验并正在评审项目建议书。

## 五、结论

文中所讨论的循环具有如下共同特点:

(1)增加功率,可使燃气轮机出力增加10%～25%;

(2)降低热耗率,可使燃气机热耗降低6%～15%;

(3)低排放,与简单循环燃气轮机相比,$NO_x$排放量降低15%～50%;

(4)投资少,每新增千瓦功率所需的投资少。

# 参 考 文 献

[1] UJI S. Partial regenerative steam injected gas turbine[C]//ASME 1996 International Gas

Turbine and Aeroengine Congress and Exhibition, GT. 1996, 4.

[2] 闻雪友,陆犇. 部分回热回注蒸汽燃气轮机循环的研究[J]. 工程热物理学报,1990, 20(3): 270 -273.

[3] DE BIASI V. Air injected power augmentation validated by Fr 7 FA peaker tests[J]. Gas Turbine World, 2002, 32(2): 12 -15.

[4] DEBIASI V. Fr7B peakers are prime prospects for humid air injection technology[J]. Gas Turbine World, 2003, 33(4): 40 -43.

# 部分回热回注蒸汽燃气轮机循环的研究

闻雪友　陆　犇

（中国船舶重工集团公司第七〇三研究所）

## 一、引言

自从美国国际动力技术公司的程大猷先生于 1976—1981 年间提出"双工质平行 – 复合循环热机"发明专利后，这种回注蒸汽燃气轮机（STIG）循环已在数个国家的多种型号的燃气轮机装置上得到应用，并取得实效。

众所周知，回热燃气轮机具有较高热效率，而 STIG 循环具有高比功。将回热燃气轮机循环与 STIG 循环结合起来，形成一种新的复合循环——回热注蒸汽燃气轮机（RSTIG）循环。RSTIG 充分利用排气余热，进一步提高了循环热效率，增大了比功，并减少处理水的消耗量。但应当指出，RSTIG 循环存在由于回热所带来的固有缺陷，例如由于大量管道的存在，导致整个系统的动态响应能力下降，同时也产生较大压损，直接影响到整个循环的热效率。因此，可以考虑将回热量减小，采用"部分回热"，也就是部分回热注蒸汽燃气轮机（PRSTIG）循环。

## 二、部分回热注蒸汽燃气轮机循环原理

典型的 PRSTIG 循环系统，如图 1 所示。

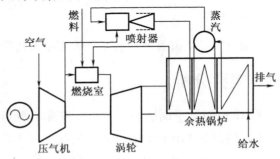

**图 1　PRSTIG 循环系统示意图**

---

\* 文章发表时间：1999 年 5 月。

一种工质空气经压气机后，分为两部分：一部分空气直接进入燃烧室；另一部分空气进入喷射器，经蒸汽引射后与之混合，形成的混合气体在余热锅炉内被涡轮排气加热，然后注入燃烧室。另一种工质（水）进入余热锅炉，从涡轮排气吸收热量，水被加热成饱和蒸汽，然后进入喷射器与空气混合。从燃烧室出来的燃气进入涡轮做功后，涡轮排气在余热锅炉内首先加热蒸汽与空气的混合气体，然后用于产生蒸汽。

## 三、PRSTIG 循环的特点

PRSTIG 循环兼具 STIG 循环和 RSTIG 循环的优缺点。

### （一）高热效率

在 STIG 循环的高热效率基础上，压气机排气的一部分经过回热后注入燃烧室，更充分利用排气余热，进一步提高循环热效率。

### （二）适中的产汽量

余热锅炉产生回注蒸汽需要大量高质量的除盐水。在 PRSTIG 循环中，涡轮排气余热首先用于回热，减少了余热锅炉的产汽量，随蒸汽排入大气的热量减少，除盐水的消耗量降低；同时，回注燃烧室内的蒸汽量减少，降低了涡轮的质量流量，STIG 循环中压气机出现喘振的可能性减小。

### （三）动态响应能力改善

在 RSTIG 循环中，当突然卸载时，即使注气阀立即关闭，由于回热器内积累了大量热能，使得空气进入涡轮后仍具有较高温度，易发生超速现象。而在 PRSTIG 循环中，当注蒸汽阀关闭后，喷射器已无法引射空气，空气自动流入燃烧室而不流经回热器，排除了超速现象。

### （四）回热器内管道量较少

回热的空气量约为 RSTIG 循环中回热空气量的一半，管道总量大幅减少。

### （五）适用的压比范围增大

PRSTIG 循环的压比适用范围基本覆盖了目前在役的工业重型燃气轮机的范围。

## 四、PRSTIG 循环简要分析

为了更具体地说明 PRSTIG 循环，下面引用一个分析结果。分析中所用的主要参数值（假定值）列于表1，使用作者编制的 PRSTIG 循环及变工况计算程序进行计算。研究了某些参数在一定范围内变化的影响，此时其他参数值一般保持常数。

表 1  分析中所用的假定值

| 序 号 | 名 称 | 数 值 |
| --- | --- | --- |
| 1 | 环境温度/℃ | 20 |
| 2 | 环境压力/bar[①] | 1.013 25 |
| 3 | 给水温度/℃ | 20 |
| 4 | 燃料注入温度/℃ | 25 |
| 5 | 压气机效率 | 0.88 |
| 7 | 涡轮效率 | 0.90 |
| 8 | 燃烧效率 | 0.99 |
| 9 | 机械效率 | 0.985 |
| 12 | 排气温度最低值/℃ | 200 |
| 13 | 回热器回热度 | 0.85 |
| 14 | 余热锅炉节点温差/℃ | 15 |

图 2 表示 PRSTIG 循环热效率与回热比 $R_a$（回热空气质量流量与压气机进口空气质量流量的比值）、回注比 $R_s$（回注蒸汽质量流量与压气机进口空气质量流量的比值）之间的关系。回注比最佳值将曲线分为上升和下降两部分：上升曲线表示此时涡轮排气中有足够的余热供余热锅炉回收，将工质加热到所要求的参数，这时空气与蒸汽组成的混合气体有较好的过热度；下降曲线表示此时回注比偏高，只好以降低注气温度（蒸汽与空气所组成的混合气体注入燃烧室时温度）作为补偿。这时混合气体在燃烧室内的吸热量增加，导致效率下降。

图 3 表示 PRSTIG 循环比功与回注比、回热比的关系。当回热比保持一定时，回注比的增加使进入涡轮的工质质量流量增加，同时单位质量的工质做功能力也增加，故比功也随之增加。当回注比保持不变时，随回热比增加，燃料注入量减小，湿燃气的平均定压比热容略有降低，导致比功略微有所下降。

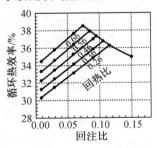 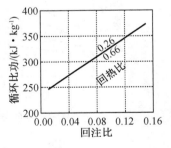

图 2  $R_a,R_s$ 对循环热效率的影响　　图 3  $R_a,R_s$ 对循环比功的影响
（燃气初温 1 273.15 K,压比 8）　　　（燃气初温 1 273.15 K,压比 8）

---

① 1 bar = 100 kPa（准确值）。

图 4 表示 PRSTIG 循环最大效率(此时回注比处于最佳值)与压比、回热比之间的关系。随着压比的升高,回热温差减小,回热比的变化对循环最大热效率的影响程度减弱。因为压比适当下降可使回热温差升高,对加热有利,故回热比越大,最佳压比越小。

图 5 表示 PRSTIG 循环最佳回注比(此时循环具有最大热效率)与压比、回热比之间的关系。

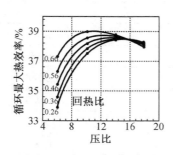 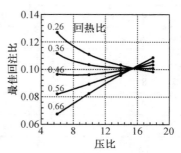

图 4  $R_a$、压比对循环最大热效率的影响　　　图 5  $R_a$、压比对循环最佳回注比的影响
(燃气初温 1 273.15 K)　　　　　　　　(燃气初温 1 273.15 K,压比 8)

图 6 表示 PRSTIG 循环性能与注气温度、回热比之间的关系。当回热比保持一定时,注气温度升高,导致机组的油气比降低,进而循环最大热效率升高;同时涡轮工质质量流量降低,并且单位质量工质做功能力下降,故循环比功略有降低。由图中还可以看出,回热比越大,注气温度对循环性能的影响越显著。

图 7 表示循环性能与燃气初温之间的关系。由图可见,燃气初温仍是影响 PRSTIG 循环性能的主要因素。

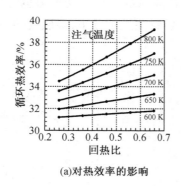 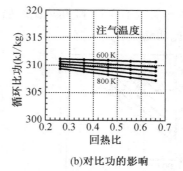

(a)对热效率的影响　　　　　　(b)对比功的影响

图 6  注气温度、回热比对循环性能的影响
(燃气初温 1 273.15 K,压比 8)

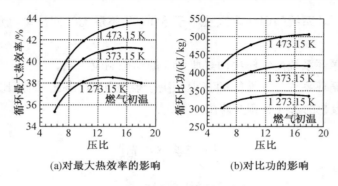

(a) 对最大热效率的影响　　　　(b) 对比功的影响

图 7　燃气初温对循环性能的影响(回热比 0.46)

综上所述,影响和决定 PRSTIG 循环性能的独立参数主要有五个:燃气初温、压比、回注比、回热比及注气温度。PRSTIG 循环性能的获得取决于这五个参数的匹配。

## 五、STIG 循环、RSTIG 循环、PRSTIG 循环性能比较

为了更加直观地说明 PRSTIG 循环的特点,将 STIG 循环、RSTIG 循环、PRSTIG 循环性能加以简单对比。计算参数选自表 1,同时取燃气初温 1 273.15 K,回热比 0.46。

图 8 表示 STIG 循环、RSTIG 循环、PRSTIG 循环最大热效率与压比之间的关系。由图中可以看出:每种循环均存在一最佳压比,此时循环具有最大热效率的最佳值,并且最佳压比大小关系为 STIG > PRSTIG > RSTIG。在小压比范围内,循环最大热效率的变化符合图 9 所示规律。由于回热的影响,当压比超过某一值 $\pi_1$ 后,RSTIG 循环的最大热效率低于 PRSTIG 循环的最大热效率;压比继续升高到另一值 $\pi_2$ 后,PRSTIG 循环的最大热效率低于 STIG 循环的最大热效率。

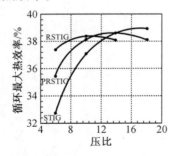　　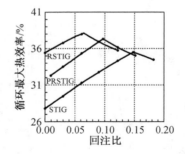

图 8　压比对循环最大热效率的影响　　图 9　回注比对循环最大热效率的影响(压比 8)

## 六、结论

(1) 压比在 12～14 范围内,PRSTIG 循环系统的热效率大于 STIG 循环系统热效率的可能性可以实现。可以考虑在 PRSTIG 循环中采用压气机间冷,以降低压气机出口空气温度,扩大 PRSTIG 循环压比适用范围,进一步提高效率。

(2) PRSTIG 循环系统的回注蒸汽量小于 STIG 循环系统的回注蒸汽量,可节省除盐水的消耗量。

(3) PRSTIG 循环系统能够解决回热所带来的动态响应问题。

(4) PRSTIG 循环系统最佳性能的获得取决于燃气初温、压比、回注比、回热比、回注气体温度五个独立参数的选择和相互匹配。

## 参 考 文 献

[1]　UJI S. Partial regenerative steam injected gas turbine[C]//ASME 1996 International Gas Turbine and Aeroengine Congress and Exhibition, GT. 1996, 4.

[2]　闻雪友. 双工质平行 - 复合循环热机(一)(程氏循环热机)[J]. 热能动力工程,1986,1(4):2 - 7,55.

[3]　闻雪友. 双工质平行 - 复合循环热机(程氏循环热机)Ⅱ[J]. 热能动力工程,1986,1(5):1 - 8,55.

# 应用于 PRSTIG 循环化 SIA – 02 燃气轮机组上的喷射器*

闻雪友 陆 犇

(中国船舶重工集团公司第七〇三研究所)

**摘 要**：喷射器是 PRSTIG 循环系统的重要组成部分。文中针对 SIA – 02 小燃气轮机组设计了喷射器，并进行了分析。

## 一、引言

部分回热蒸汽燃气轮机循环[1]是 STIG 循环的一种变异，典型的 PRSTIG 循环系统如图 1 所示。

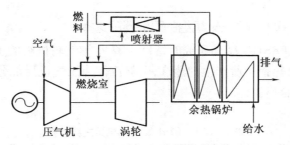

**图 1 PRSTIG 循环系统示意图**

一种工质空气经压气机后，分为两部分：一部分空气直接进入燃烧室；另一部分空气进入喷射器，经蒸汽引射后与之混合，形成的混合气体在余热锅炉内被涡轮排气加热，然后注入燃烧室。另一种工质(水)进入余热锅炉，从涡轮排气吸收热量，水被加热成饱和蒸汽，然后进入喷射器与空气混合。从燃烧室出来的燃气进入涡轮做功后，涡轮排气在余热锅炉内首先加热蒸汽与空气的混合气体，然后用于产生蒸汽。

PRSTIG 循环借部分回热进一步提高了循环热效率，又减少了回注蒸汽所需高质量除盐水的消耗。由于采用喷射器，也使 PRSTIG 循环的动态响应能力改善。

喷射器是此循环系统中的重要装置，直接影响了 PRSTIG 循环系统的性能。在喷射器

---

* 文章发表时间：2000 年 5 月。

中,具有不同压力、温度的压缩空气(引射介质)和蒸汽(工作介质)相互混合,发生能量交换,达到压力和温度的均衡,产生一定压力和温度的混合气体。七〇三所拟在已成功进行了 STIG 循环试验的 SIA – 02 小燃气轮机组上,进行 PRSTIG 循环研究。文中即是对其喷射器进行分析。

## 二、喷射器的选择

### (一)喷射器的分类

喷射器的工作情况取决于相互作用介质的相态和弹性特性。PRSTIG 循环中,蒸汽和空气均为气态,因此,气体的弹性特性决定了喷射器的工作过程;而所谓弹性特性是指介质的比体积随压力的改变而改变的特性。

对于弹性介质的同相喷射器,其工作取决于引射介质的压缩比(喷射器扩散器后压缩流体压力与喷射器接收室前引射流体压力的比值)和工作介质的膨胀比(喷射器喷嘴前工作流体压力与喷射器接收室前引射流体压力的比值)。由此,喷射器可分为以下三类:

(1) 大膨胀比和大压缩比喷射器 – 气体引射器,通常这类喷射器所能建立的压缩比大于 2.5;

(2) 大膨胀比和中压缩比喷射器 – 气体喷射压缩器,通常这类喷射器所能建立的压缩比在 2.5 和 1.2 之间;

(3) 大膨胀比和小压缩比喷射器 – 气体喷射器,通常这类喷射器所能建立的压缩比小于 1.2。

### (二)适用于 SIA – 02 小燃气轮机组的喷射器的选择

表 1 列出了进行 PRSTIG 循环研究时,SIA – 02 小燃气轮机组的主要参数。在 PRSTIG 循环中,工作流体为余热锅炉所产生的蒸汽,引射流体为压气机后部分压缩空气,压缩流体为蒸汽与压缩空气混合后流体。

**表 1  SIA – 02 小燃气轮机组主要参数**

| 参　数 | 数　值 | 参　数 | 数　值 |
|---|---|---|---|
| 涡轮入口温度/℃ | 813.85 | 压气机入口温度/℃ | 15.00 |
| 压气机入口压力/MPa | 0.101 | 压气机入口空气流量/(kg/s) | 1.676 |
| 压气机压比 | 9.20 | 蒸汽回注比 | 0.072 |
| 喷注蒸汽温度/℃ | 413.00 | 回热度 | 0.85 |
| 蒸汽压力/MPa | 1.20 | 给水温度/℃ | 20.00 |
| 压气机效率 | 0.757 6 | 涡轮效率 | 0.815 3 |
| 注气压力/MPa | 0.935 1 | 空气回热比 | 0.460 |

由膨胀比和压缩比的情况可知,只有气体喷射器适用于 SIA – 02 机组的 PRSTIG 循环研究。

## 三、喷射器的优化设计

气体喷射器尺寸的计算[2]主要依据 PRSTIG 循环系统对气体喷射器的性能要求和工作流体、引射流体的初始条件。一般地,优化设计的目的是使气体喷射器的出口具有尽可能高的压力。以下依据表 1 所列的参数,对喷射器进行优化设计。

由于引射流体与工作流体的压力比大于工作流体的临界压比,工作喷嘴出口截面上工作流体的速度小于临界速度,于是采用锥形喷嘴。喷射器其余部分的轴向、径向尺寸计算较为复杂,一般采用经验公式[2]。

喷射器的最佳截面比对应喷射器所能形成的最大压力升值。

$$\frac{d\Delta(P_c)}{d(f_{p1}/f_3)} = 0$$

式中　$\Delta P_c$——喷射器出口压力升值;

　　　$f_{p1}$——锥形喷嘴出口截面积;

　　　$f_3$——混合室入口截面积。

应用于 SIA - 02 机组的喷射器优化设计计算结果如图 2 所示。

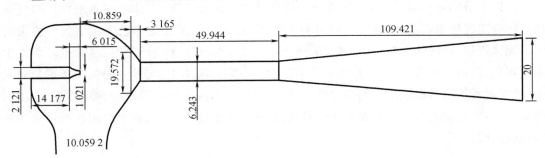

**图 2　喷射器尺寸图**(单位:cm)

## 四、喷射器的变工况分析

在实际的 PRSTIG 循环系统应用中,发生变化的参数一般只是回注比、回热比,因此对已设计完成的应用于 SIA - 02 燃气轮机组的气体喷射器进行变工况分析是必要的。气体喷射器如表 1、图 2 所示。

图 3、图 4 分别为回热比(回热空气质量流量与压气机进口空气质量流量之比)$rbc$、回注比(回注蒸汽质量流量与压气机进口空气质量流量之比)$g_s$ 对 $d_p$($d_p$ = (混合流体压力 - 引射流体压力)/引射流体压力)和 $T_c$(喷射器出口混合气体温度)的影响,图中 $c$ 点表示设计状态点。

图 3 中,在 $rbc$ 不变,$g_s$ 大于设计回注比的情况下,$d_p$ 随 $g_s$ 的增加而增加,升幅较缓;在 $g_s$ 小于设计回注比的情况下,$d_p$ 随 $g_s$ 的减小而急剧下降,甚至出现 $d_p$ 小于零的情况,即经

气体喷射器引射后,混合气体的压力低于压气机出口的压缩空气的压力。在 $g_s$ 不变时,$d_p$ 随 $rbc$ 的增加而下降。$g_s$ 小于设计状态点,$rbc$ 对 $d_p$ 的影响程度急剧上升。图4中,在 $rbc$ 不变时,$g_s$ 增大,$T_c$ 下降;在 $g_s$ 不变时,$rbc$ 增大,$T_c$ 上升。并且在功率输出不变时,无论 $g_s$、$rbc$ 怎样偏离设计状态点,对 $T_c$ 的影响程度不大。

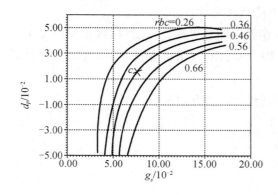

图3　回注比、回热比对压力的影响　　　图4　回注比、回热比对温度的影响

由以上分析可以看出,在回注比小于设计状态点的情况下,气体喷射器出口压缩气体压力的变工况性能较差,此时气体喷射器发生了阻塞现象,工作流体已无法引射所要求的引射流体流量,且回注比越小,阻塞现象越严重。而回热比大于设计值时,尽管也可能发生阻塞现象,但由图中可以看出,其影响并不很大。在实际应用中,应尽量避免阻塞现象的发生,使喷射器的出口参数能达到最佳值。而正是由此,当注蒸气阀关闭后,喷射器已无法引射空气,空气自动流入燃烧室而不流经回热器,排除了超速现象,能够解决回热所带来的动态响应问题。

## 五、结论

(1)为了满足 PRSTIG 循环对喷射器的要求,采用气体喷射器较为适宜。

(2)对应用于 SIA-02 PRSTIG 试验装置的气体喷射器进行了优化设计,建立了回热比、回注比、气体喷射器出口温度、气体温度和压力间的关联。

(3)喷射器的变工况分析表明,对于某一已设计完成的 PRSTIG 系统存在一个最小回注比,因此 PRSTIG 循环更适用于长期以稳定工况工作的机组。

## 参 考 文 献

[1] UJI S. Partial regenerative steam injected gas turbine[C]//ASME 1996 International Gas Turbine and Aeroengine Congress and Exhibition, GT. 1996, 4.

[2] 索科洛夫 Е Я,津格尔 Н М.喷射器[M].黄秋云,译.北京:科学出版社,1977.

# 柴油机注汽涡轮增压系统[*]

(闻雪友,陆 犇,夏军宏 中国船舶重工集团公司第七〇三研究所)
(张玉龙 哈尔滨汽轮机有限责任公司军代表室)

**摘 要**:针对涡轮增压柴油机在低工况运行时,出现增压压力不足、燃烧过量空气系数小和废气排温较高等固有特性,提出了一种利用增压器废气余热产生水蒸气,并注入涡轮来提高增压器的压比和空气量的新方法,以改善涡轮增压器与柴油机的匹配,提高柴油机性能。

## 一、引言

随着涡轮增压技术的发展,柴油机的功率、经济性、质量和体积等各项性能指标都获得了明显的改善。现代大功率柴油机几乎全部采用了涡轮增压装置,而其中应用最广泛的是废气涡轮增压装置。

涡轮增压柴油机是柴油机与涡轮增压器所组成的复合式发动机。在各种工况下,涡轮增压装置应该向柴油机提供足够的空气量,如果涡轮增压器的供气量不足,就会引起工作过程的亚化。然而,当柴油机工况变化时,废流量和参数也都发生变化,往复式的柴油机与回转式涡轮增压器两者特性匹配的结果通常是涡轮增压柴油机在低工况运行时会出现增压压力不足,燃烧过量空气系数小和废气排温较高等固有特性,其直接表现就是涡轮增压器不能提供运行条件所要求的压比,以满足对空气量的要求。

图1表示涡轮增压柴油机在不同运行条件下的涡轮增压器的运行特性(一个特定的涡轮增压器的运行特性),即在不同的涡轮进口折合流量下所能提供的压比。由图可见,在标定工况(即设计点),增压器提供的压比与柴油机要求的相符;在低工况时,运行特性不同,提供与要求的压比差别不同,按车用特性($Ne = cn^2$、$Ne = cn^3$($Ne$ 为功率,$n$ 为转速))运行时增压器所能提供的压比均不能满足要求。

为了改善涡轮增压柴油机的低工况性能,当前在涡轮增压系统上采用了多种措施,各种改进方案也均有其各自的优缺点。

---

[*] 文章发表时间:2003年3月。

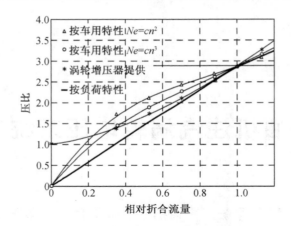

图1  涡轮增压器配合性能

## 二、系统简介

文中提出一种新型涡轮增压器系统,其能大幅度提高涡轮增压器的压比,尤其是低工况下的压比,从而改善低工况的性能,提高涡轮增压内燃机的功率,降低涡轮增压内燃机的耗油率。

本系统是在涡轮增压器的基础上,利用涡轮增压器的排气余热,设置一个朗肯回热循环,构成一种新颖的注汽涡轮增压系统。借余热所产生的蒸汽注入涡轮中将大幅度地提高涡轮功率,使由其驱动的压气机的压比及流量增加,从而达到提高涡轮增压柴油机的功率,降低其耗油率,明显改善低工况性能的目的。

本系统原理图如图2所示。其主要流程是:工质空气经压气机1增压进入内燃机气缸2,压缩燃烧做功后排气进入涡轮3。水作为第二种工质进入余热锅炉4,在余热锅炉4内从涡轮排气余热中吸收热量变为过热蒸汽,通过专门的注汽系统注入涡轮前。在涡轮3中由两种工质(燃气、蒸汽)组成的混合工质膨胀做功驱动压气机,最后经余热锅炉4排入大气。回注的蒸汽量可由调节阀5控制,来满足涡轮增压内燃机的需要。多余蒸汽可由旁路系统6引作他用或排放。系统8为涡轮排气旁路调节系统。

### (一)注汽涡轮增压器

涡轮增压器中有两个独立变量:涡轮进口(内燃机废气出口)的燃气温度和压力。但是在注汽涡轮增压器中还有一个独立变量:蒸汽质量流量与燃气质量流量之比,即汽气比。汽气比的大小对涡轮增压器的压比、转速的变化影响较大,因为在等压下蒸汽比热容至少要比空气－燃料燃烧产物的比热容高一倍,因此蒸汽的做功能力更大。注汽温度对压比、转速的影响则相对较小。图3给出一特定增压器注汽后特性的变化,随着汽气比增大,涡轮增压器的压比增加。

当选择汽气比时应综合考虑如下因素作出最佳选择:

(1) 内燃机对增压器压比、流量的要求;
(2) 压气机的喘振裕度;
(3) 涡轮增压器的转速极限;
(4) 涡轮的通流能力;
(5) 余热锅炉的紧凑性(尺寸、质量);
(6) 兼顾全回注与零回注。

注汽涡轮增压器的背压因涡轮出口装有余热锅炉而升高,但是注汽后获得的功率增益远大于背压升高而带来的损失。

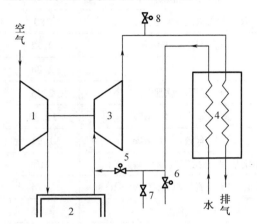

图2 注蒸汽的涡轮增压柴油机示意图

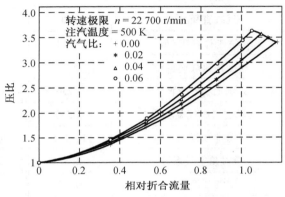

图3 汽气比对增压器压比的影响

(二) 余热锅炉

余热锅炉吸收涡轮增压器排气的热量产生蒸汽。在涡轮增压器以非回流方式运行时,涡轮排气的成分是燃气;在以回注方式运行时,涡轮排气成分是燃气加蒸汽。

涡轮增压柴油机系统要求余热锅炉结构高度紧凑、质量轻、热惯量小。其设计思想不是最大限度地利用排气余热多产蒸汽,而是根据涡轮增压器的要求,在蒸汽产量和余热锅炉尺寸、质量间作折中。余热锅炉结构宜选择直流锅炉或强制循环型,采用强化传热元件,或高度紧凑的特种结构。余热锅炉应设计成允许干烧,当余热锅炉某些部件发生故障或要求做零注汽运行时,余热锅炉可以做干式运行。

(三) 注汽系统

为了将过热蒸汽从余热锅炉导入涡轮增压器,需设置回注系统。

系统主要包括注汽流量调节阀、注汽喷嘴、吹扫系统、蒸汽分路系统。图2中,吹扫系统7为注汽管系在工作前预热,防止有凝水进入增压涡轮;旁路系统6为排放可能的多余蒸汽以及在柴油机停机,而余热锅炉因热惯量较大,继续产生蒸汽的排放之用。

## 三、系统性能

为了进一步说明注汽涡轮增压系统的优越性,引入一个示例。

某涡轮增压柴油机的标定参数见表1。

表1　涡轮增压柴油机的标定参数

| 标定参数 | 数值 | 标定参数 | 数值 |
| --- | --- | --- | --- |
| 气缸数 | 12 | 气缸排气温度/℃ | 450 |
| 活塞行程/mm | 460 | 涡轮增压器后燃气温度/℃ | 380 |
| 气缸直径/mm | 400 | 标定压气机效率 | 0.72 |
| 单缸工作容积/dm$^3$ | 57.81 | 标定压气机压比 | 2.88 |
| 额定转速/(r/min) | 520 | 标定涡轮膨胀比 | 2.45 |
| 额定功率/kW | 5 736 | 最大爆发压力/MPa | 12.35 |
| 涡轮进口燃气温度/℃ | ≤580 | 平均有效压力/MPa | 1.91 |
| 涡轮排气背压/kPa | 1~1.5 | 标定燃油消耗率/(g/(kW·h)) | 210 |
| 总过量空气系数 | 2.32 | 标定进空气流量/(kg/(kW·h)) | 7.14 |
| 空气利用系数 | 0.48 | 进空气温度/℃ | ≤45 |
| 涡轮增压器总效率 | 0.55 | 喷嘴环当量面积/cm$^2$ | 177 |

图1已示出该涡轮增压柴油机在不同运行条件下的涡轮增压器配合性能,并已指出,在中、低工况时,增压器所能提供的压比并不能满足按车用特性 $Ne = cn^2$、$Ne = cn^3$ 运行对压比的要求,现用注汽涡轮增压柴油机系统作为改善措施。

图3为采用注汽涡轮增压系统后,该增压器在不同的汽气比下所能获得的压比增加值。

图4为采用注汽涡轮增压系统后涡轮增压器的运行特性,该特性是在汽气比为0.06下获得的。由图4可见,注蒸汽后的涡轮增压器已能满足 $Ne = cn^2$,$Ne = cn^3$ 对压比的要求,即使按车用特性运行时,其性能也大为改善,并且在标定工况和高工况时,在涡轮增压器许用转速内,也可使压比有较大升高。

图5则最终反映了注汽涡轮增压柴油机系统的运行效果,耗油率平均下降约1.8%(图5(a)),柴油机的功率平均增加约16%(图5(b))。注汽前后涡轮增压柴油机的功率和耗油率的改进是明显的。

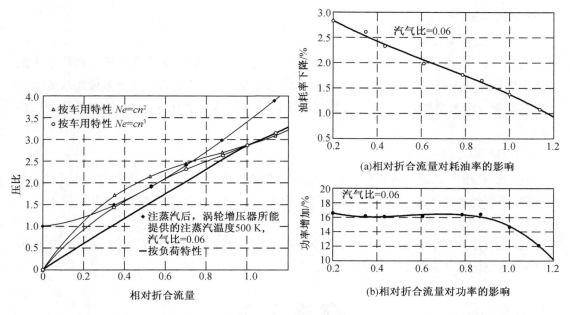

图 4　注蒸汽后的涡轮增压器配合性能

图 5　汽气比 0.06 时,相对折合流量对涡轮增压柴油机的性能变化的影响

## 四、系统特点

### (一)显著提高柴油机功率

由于第二工质为水,在等压下的蒸汽比热容至少要比空气－燃料燃烧产物的比热容高一倍。换言之,蒸汽作为主工质能比内燃机排气(空气－燃料燃烧产物)做更大的机械功。随着汽气比的增加,增压涡轮的功率随之增大,从而使压气机的流量、压比增加,最终使内燃机的功率大幅度增加。

### (二)提高柴油机效率

相对于原增压系统而言,新系统增加了余热锅炉以产生蒸汽,这增加了泵水的功耗。其实,泵水加压的功耗不大,但将水变为蒸汽则需大量的能耗,而这在本系统中是需要利用排气余热实现的,这就改善了注汽涡轮增压系统的效率,从而提高了柴油机的效率。

### (三)独特的参数匹配和调节方式

通常的涡轮增压器设计中有两个独立变量:涡轮进口温度和压比。注汽涡轮增压系统设计中有三个独立变量:涡轮进口温度、压力和汽气比。对于一个已与柴油机配装的涡轮增压器而言,涡轮的进口参数即是由柴油机的排气参数所确定;而对于一个与柴油机配装的注汽涡轮增压器而言,其涡轮进口参数仍可通过汽气比的变化来加以调节。采用注汽涡轮增压器后,增压器出口的空气压力、流量在各工况下都有大幅度提高,加之有注汽调节系统和涡轮排气旁路调节系统,综合运用可以获得一个与运行规律所要求的涡轮增压器特性相逼近的特性。

### （四）部分负荷性能良好

通常的涡轮增压器在部分负荷时因涡轮驱动功率不足,而使增压器的压比和流量未能达到内燃机在此工况下的期望值。在注汽涡轮增压系统中,涡轮的质量流量可用改变蒸汽流量的简单方法来增加涡轮的输出,提高增压器的空气流量和压比,因而使部分负荷性能良好。

### （五）现实可行性

注汽涡轮增压系统可以在原有的涡轮增压器上改装实施,主要是附加了一些设备、系统,而对原涡轮增压器的改动较小,所获得的内燃机功率和耗油率的改善又很明显,是一个现实可行的方案。

实际上,注汽后增压涡轮进口的混合工质的工作温度明显低于原来未注汽的工作温度,因此这也使增压涡轮的工作寿命延长。

## 参 考 文 献

[1] 闻雪友,陆犇. 内燃机注汽涡轮增压系统：CN,1363764[P]. 2002-08-14.
[2] 闻雪友. 双工质平行－复合循环热机（程氏循环热机）Ⅱ[J]. 热能动力工程,1986, 1(5):1-8,55.
[3] 顾宏中. 涡轮增压柴油机性能研究[M]. 上海：上海交通大学出版社,1998.
[4] 闻雪友. 双工质平行－复合循环热机（一）（程氏循环热机）[J]. 热能动力工程,1986, 1(4):2-7,55.

# PG5361 STIG 装置*

闻雪友　魏应新
(哈尔滨船舶锅炉涡轮机研究所)

**摘　要**:文中简述了我国第一套自行研制的电站用 STIG 装置的情况。该装置回注蒸汽后在功率、耗油率方面的得益令人印象深刻。

## 一、引言

自国际第一套双工质平行－复合循环,即回注蒸汽轮机循环(STIG)商用装置于1985年正式投入商业运行以来,短短数年间,这种 STIG 装置获得了相当迅速的发展。它的特点是:高比功,高效率,热、电负荷平衡方面具有高度的灵活性,排放物中的 $NO_x$ 含量减少,投资费用低,可由已有的机组改造。

STIG 在国外主要用于热电联供装置和发电装置,是20世纪80年代国际上该领域的热点研究项目之一。

深圳南山热电有限公司有三台 ALSTHOM 公司制造的 PG5361 P1 型燃气轮机发电机组,由于深圳气候炎热和机组老化,机组常年在远低于铭牌功率的条件下运行,周围又有热用户,因此将机组改造为 STIG 装置实现热电联供可明显提高营运效益。由于哈尔滨船舶锅炉涡轮机研究所具有在1984—1989年间的研究工作及已建成 STIG 整机试验台并完成试验的基础,上述建议被采纳。并由该所总承包。

## 二、STIG 装置

PG5361 STIG 装置的工作原理如图1所示。

----

* 文章发表时间:1992年7月。

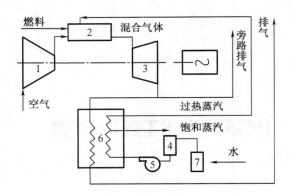

**图 1　PG5361 STIG 装置的工作原理图**
1—压气机；2—燃烧室；3—涡轮；4—除氧水箱；5—水泵；6—余热锅炉；7—水处理装置

空气经压气机后进入燃烧室，燃料喷入燃烧室燃烧。第二种工质水则在余热锅炉中吸收涡轮排气（燃气、水蒸气的混合物）的废热变为蒸汽。一部分饱和蒸汽从余热锅炉中引出供热，一部分饱和蒸汽经过热器后成为过热蒸汽，通过专门的回注系统回注到燃烧室中，与燃烧后的高温气体混合进入涡轮中膨胀做功后经余热锅炉排入大气。当然，通过三通烟道挡板的切换，燃气轮机仍可以简单循环方式运行。

三套装置的供热系统的母管是连通的，注汽系统是各自独立的。根据电厂要求，系统以供热为主，多余的蒸汽回注，增加发电量，降低油耗，即回注作为调节热负荷的手段，充分发挥了 STIG 循环热电负荷可灵活匹配的特点。

（一）燃气轮机发电机组

基础发动机是法国阿尔斯通公司生产的屋外集箱式 PG5361 P1 型燃气轮机。在 ISO 条件下性能如下：

机组出力为 24 690 kW（烧轻油）；

吸入空气量为 436 000 kg/h；

燃油量为 7 600 kg/h（烧轻油）；

排出烟气量为 439 000 kg/h；

排烟温度为 493 ℃；

机组热耗率为 13 190 kJ/(kW·h)；

转速为 5 100 r/min。

发电机形式为开口通风冷却系统，型号为 T174 - 160，其性能如下：

出力为 30 862 kW；

转速为 3 000 r/min(50 Hz)；

功率因数为 0.8；

出线电压为 11 000 V。

（二）余热锅炉

余热锅炉为无补燃双压自然循环形式，高压蒸汽供回注和供热，低压蒸汽供除氧用，其原理如图 2 所示。来自燃气轮机的排气依次冲刷过热器、高压蒸发管束、省煤器、低压锅炉蒸发管束,最后排入大气中。

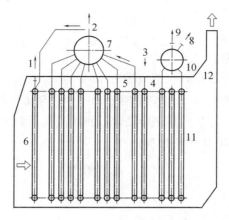

**图 2　余热锅炉原理图**

1—过热蒸汽出口;2—饱和蒸汽出口;3—高压给水;4—省煤器;5—蒸发器;6—过热器;7—高压汽包;
8—低压给水;9—低压蒸汽出口;10—低压汽包;11—低压蒸发器;12—烟囱

余热锅炉设计性能为:当燃气轮机烧重油、在 STIG 循环回注蒸汽量 7%(蒸汽量与空气量之比)、烟气流量 361 083 m$^3$/h、进口烟温 469 ℃、给水温度 104 ℃、高压蒸汽压力 1.86 MPa、过热蒸汽温度 280 ℃时,蒸汽产生量为 50 t/h(其中过热蒸汽 30 t/h),低压蒸汽压力为 0.2 MPa,产量为 8 t/h。

受热面采用双集箱先进高效传热的螺旋圈片组合元件,组合元件由螺旋圈片管和上、下集箱组成。

余热锅炉用水为自来水,经一级除盐和除氧后供给锅炉。

(三)烟道

燃气轮机余热回收装置中有一套尺寸相当大的烟道系统和相应的配套设备,包括主烟囱、旁通烟囱、三通烟道、过渡段、热膨胀补偿元件和烟气挡板门及钢架等。本套装置主烟囱高 30 m,旁通烟囱高 25 m。热膨胀补偿元件采用单波金属膨胀节、双波金属膨胀节及非金属膨胀节,保证系统可靠运行。新研制的烟气挡板门采用矩形高温电动蝶阀,分别装于主烟道与旁通烟道上,阀体及阀瓣采用耐热铸铁制造,阀体与阀瓣之间有密封结构。

(四)回注蒸汽系统

过热蒸汽自锅炉过热器出来,经减温、蒸汽过滤器,进入蒸汽回注撬体。蒸汽再经过气动速关阀、旋涡流量计、电动蒸汽调节阀、止回阀、蒸汽分汽管、蒸汽环管,并经金属软管连接到 10 个燃烧室各自的小环管上,最后通过蒸汽喷嘴注入燃烧室。蒸汽、二次空气与燃烧后的高温燃气混合进入涡轮膨胀做功,其流程如图 3 所示。

该系统可分为两部分,在蒸汽滤以后止回阀以前的管路和设备装于蒸汽撬体内。止回阀以后设备有蒸汽分配及注汽部分布置在燃机箱体内。整个回注系统的管路、预热、吹扫管线以及有关阀门蒸汽的控制等,都是由计算机自动控制的,在满足了回注的诸要素后即可开始回注蒸汽。

回注蒸汽参数如下:

蒸汽压力为 1.67 MPa;

蒸汽温度为 280±30 ℃;

蒸汽流量为大于或等于 5 t/h,小于或等于 30 t/h。

图 4 是一张实测的性能曲线,试验时的条件如下:平均大气温度 23 ℃,机组以简单循环运行的输出功率为 19 MW,当以 STIG 循环运行,回注蒸汽量为 7% 时,其实测性能与简单循环比较,功率提高 30%,热耗率下降 15%。

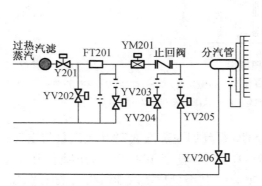

图 3  回注蒸汽系统图

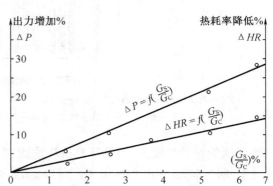

图 4  燃机出力增加、热耗率降低与回注量的变化关系

## (五)调控监测系统

整个装置的调控监测系统由三部分组成。

### 1. 燃气轮机调控监测系统

原三台燃气轮机发电机组设有一个主控室,可监测一些主要运行参数,而燃气轮机调控和监测是以每台机为单元控制,设在机头就地控制室内。燃机的主要控制系统有启动控制、速度控制、温度控制和加速控制。通过一次元件可监测燃机的转速、温度和压气机出口压力。保护设有温度、转速、振动和火焰监测等保护系统。

### 2. 余热锅炉调控监测系统

可以在就地人员的配合下,在集控室内进行余热锅炉的启停,并进行参数的自动调节及正常运行的监控。三台余热锅炉共有 10 个自动调节回路,可进行自动控制或遥控,分别是主蒸汽压力(供热蒸汽总管压力,三台共用)、过热蒸汽温度(每炉各一)、高压锅炉水位(每炉各一)、低压锅炉水位(每炉各一)等自动调节系统。余热锅炉吹灰可实现程序控制。对锅炉给水、排污、放水等可遥控。控制室可对锅炉蒸汽压力、温度、流量、给水压力、汽包水位、给水流量、烟气侧压力、温度等进行监测。

余热锅炉、回注蒸汽系统、给水除氧等控制设在同一集中控制室内。余热锅炉、回注蒸汽系统主要热工参数可通过计算机进行模拟图显示、表格打印和将主要运行参数存入磁盘中。

## 三、运行模式

简单循环燃气轮机改为 STIG 循环并热电联供时,运行模式具有非常大的灵活性。燃气轮机既可按简单循环模式运行,也可按复合循环模式运行。复合循环既可按全供热模式运行(相当 0% 回注),也可按只回注不供热模式运行(相当供热为 0)。当三台炉均工作时,最大蒸汽供热能力为 150 t/h,供热蒸汽量的调节范围大于或等于 60 t/h 而小于 145 t/h,回注蒸汽量的调节范围大于 5 t/h 而小于或等于 90 t/h。可见热、电匹配十分灵活。需指出的是,运行模式的转换和供热与回注蒸汽的匹配都是在集控室内由 1~2 人操作的,且后者是自动调节的。当然,在启动时,尚需少量人在现场配合。

## 四、结论

从以上 STIG 技术应用实例中,不难看出,对现有燃气轮机发电机组进行改造有着很大的运行经济效益:

(1) 可以使由于环境温度升高、通流部分结垢及长期运行后出现的机组功率下降得到恢复和提高;

(2) 可以使热耗率下降,直接提高经济效益;

(3) 可以降低机组运行初温,从而大大延长机组运行寿命,间接地提高了经济效益。

实践证明,STIG 技术装置构成简单,占地面积小,而且投资费用低,一般来说投资回收期只需一年多的时间。

这一技术不仅适用于 MS 5000 系列机组,也适用于一切能进行回注蒸汽的燃气轮机机组。

## 参 考 文 献

[1] 闻雪友. 双工质平行 - 复合循环热机(一)(程氏循环热机)[J]. 热能动力工程,1986, 1(4):2 - 7,55.

[2] 闻雪友. 双工质平行 - 复合循环热机(程氏循环热机)Ⅱ[J]. 热能动力工程,1986, 1(5):1 - 8,55.

[3] 闻雪友. STIG 的新进展[J]. 热能动力工程,1990,5(1):10 - 13.

[4] 邹积国,傅正. 程氏循环装置设计工况及变工况计算分析[J]. 热能动力工程,1990, 5(3):1 - 6,11.

[5] 臧述升. 用残量法预估注蒸汽燃气轮机变工况性能[J]. 热能动力工程,1990, 5(3):7 - 11.

[6] 闻雪友,金介荣,傅正,等. 燃气轮机回注蒸汽装置的研究[J]. 热能动力工程,1991, 6(3),123 - 125,131.

[7] 邹积国,闻雪友. 燃气轮机 STIG 化的研究[J]. 热能动力工程,1991,6(3):126 - 131.

# M1A – 01 – CC 程氏循环热电联供装置*

闻雪友
（哈尔滨船舶锅炉涡轮机研究所）

日本川崎重工业株式会社（KHI）自 1972 年起一直在进行工业燃气轮机的发展研究工作。KHI 最初发展 S1A – 01 型（190 kW）发电用燃气轮机，当取得预期效果后又开始发展 M1A – 01 型（1177 kW）和 S2A – 01 型（700 kW）两型燃气轮机。两型机在气动设计概念及结构方面都是类似的：简单形式循环，两级离心式压气机，单管回流式燃烧室，多级轴流式涡轮。1977 年，KHI 开始生产应急/备用燃气轮机发电机组，现在已有额定功率 180 ~ 2 400 kW 的 14 型燃气轮机发电装置，约 1 500 台燃气轮机在各应用场合作为应急和尖峰负荷使用。

几年前，川崎重工与美国国际动力技术公司（IPT）签订了制造、使用和销售程氏循环发电系统的协议，该协议允许川崎重工在全世界独享基础发动机功率（未注蒸汽的功率）在 500 ~ 2 000 kW 间的程氏循环系统的市场。

程氏循环用于热电联供系统的主要优点如下：

（1）利用把燃气轮机排气余热产生的过热蒸汽注入燃烧室的方法，提高了发动机的功率并改进了热效率；

（2）电热比值可随季节和时间的要求而变化，在满足热电平衡方面非常灵活；

（3）程氏循环可使 $NO_x$ 的排放大大减少；

（4）程氏循环简化了 COGAS 的 RACER 设计概念，因而比联合循环更紧凑，成本更低。

川崎重工首先选用 M1A – 01 发动机（图 1）来发展一个程氏循环系统。M1A – 01 压气机的压比为 8，流量为 7.7 kg/s，三级轴流涡轮，质量（干重）约为 3 t，外形尺寸为 2.2 m × 1.4 m × 1.4 m。

川崎重工的发展工作分两步进行。首先进行改型发动机的原型试验，然后建造一生产型机组安装于明石工厂进行运行验证。1986 年 2 月开始原型机试验，发动机的改动主要是在燃烧室上增加了蒸汽喷注系统（图 2），并对涡轮导向器面积做调整，以适应蒸汽喷注。在原型

---

\* 文章发表时间：1989 年 3 月。

机上是用打磨第一、二级导向叶片尾缘的方法来实现的(在生产型机上用改变导向叶片安装角及增加叶片高度来实现)。初时,齿轮箱尚未更换,因此试验必须在 1 500 kW 以下进行。1987 年春,换上新的生产型齿轮箱,使试验功率达到 1 860 kW。在原型机试验中蒸汽是由另一单独的锅炉提供的。1988 年 1 月第一套生产型装置在现场安装完毕,该装置采用多单元撬装件设计,因而安装时间短,费用低。装置经试验调整于 1988 年 4 月正式投运,发电功率提高约 70%,发电效率提高约 30%。整个装置的流程如图 3 所示,平面布置如图 4 所示。装置可借计算机控制自动运行和监测,使其一直在最佳条件下运行,并设有燃气轮机故障预告系统。

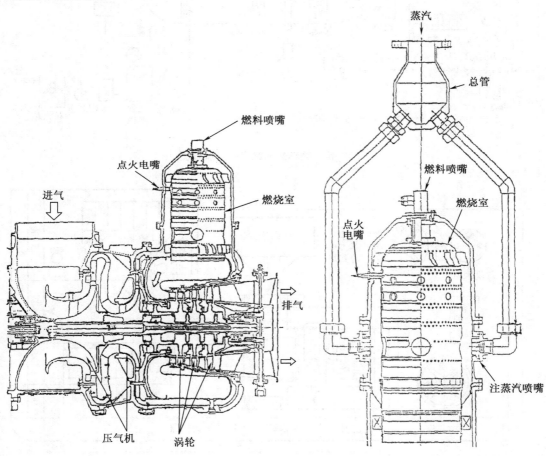

图 1  M1A-01 燃气轮机　　　　图 2  M1A-01-CC 的喷注蒸汽燃烧室

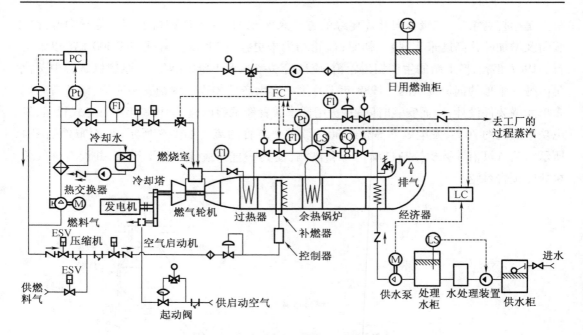

**图 3　程氏循环热电联供装置流程图**

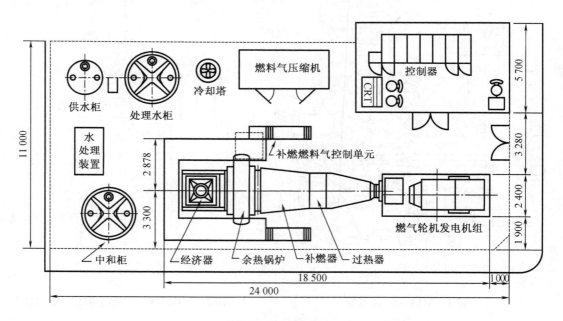

**图 4　明石工厂程氏循环热电联供装置平面布置图**

装置各部分的性能如下。

(1) 燃气轮机

型号　　　　　　　　　　M1A-01-CC

| | |
|---|---|
| 额定功率 | 1 940 kW(蒸汽注入量 1.2 kg/s) |
| | 1 090 kW(无蒸汽注入) |
| 转速 | 涡轮转子 22 000 r/min |
| | 发电机 1 800 r/min |
| 燃料 | 城市煤气(13A),煤油 |

(2)发电机

| | |
|---|---|
| 形式 | 空冷,无刷 |
| 功率 | 1 750 kW |
| 电压 | 6 600 V |
| 频率 | 60 Hz |

(3)锅炉

| | |
|---|---|
| 形式 | 自然循环水管锅炉 |
| 额定压力 | 1 422 kPa(14.5 kgf/cm$^2$) |
| 额定蒸汽产量 | 9 000 kg/h(使用补燃时) |
| 蒸汽温度 | 饱和蒸汽温度 |

(4)蒸汽过热器

| | |
|---|---|
| 额定压力 | 1 373 kPa(14 kgf/cm$^2$) |
| 额定蒸汽产量 | 4 320 kg/h |
| 蒸汽温度 | 428 ℃ |

(5)补燃器

| | |
|---|---|
| 形式 | 带火焰稳定器 |
| 燃料 | 城市煤气(13A) |
| 燃料量 | 400 m$^3$/h(最大) |
| 调节范围 | 10% ~ 100% |

(6)气体燃料压缩机

| | |
|---|---|
| 形式 | 螺杆压缩机 |
| 入口压力 | 98 kPa(1 kgf/cm$^2$) |
| 出口压力 | 1 373 kPa(14 kgf/cm$^2$) |
| 功耗 | 73 kW |

整个程氏循环热电联供装置的性能如图5和表1所示。图5清晰地表明由于采用了喷注蒸汽和补燃,大大扩大了可能的运行区域。A区域内各点的特征是有不同的蒸汽喷注量和补燃量,而燃气初温不变(为最大持续功率时的燃气初温值);B区域的特征是有不同的蒸汽喷注量和补燃量,而无补燃;C区域的特征是有不同的燃气初温和补燃量,而无蒸汽喷注。上述性能均系在ISO标准条件、气体燃料低热值为41 577 kJ/m$^3$(9 930 kcal/m$^3$)、发电机效率为95%、过程蒸汽为1 433 kPa(14.5 kgf/cm$^2$)下的饱和蒸汽、给水温度为10 ℃,进

气损失 980.6 Pa(100 mmH$_2$O)、排气损失 3 922.4 Pa(400 mmH$_2$O)条件下的值。分析表明，其最大汽气比在该燃气初温和压比下相应于最大循环效率点。

川崎现正在开发以 M1A–13 型为基础的 M1A–13CC 程氏循环系统。M1A–13 型是 M1A–01 型的发展型，其参数比 M1A–01 型有所提高，因而功率也增大。两者的结构类同，具有发动机整体互换性，但部件不能互换。M1A–13CC 的估计性能亦示于图 5（以虚线表示）和表 1。

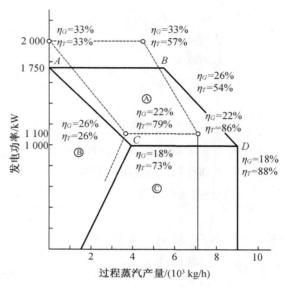

**图 5　程氏循环装置性能**

**表 1　程氏循环装置性能**

| 型 号 | M1A–01–CC 程氏循环 | | | | M1A–13CC 程氏循环 | | | |
|---|---|---|---|---|---|---|---|---|
| 运行点 | A | B | C | D | A | B | C | D |
| 输出功率/kW | 1 750 | 1 750 | 988 | 988 | 2 000 | 2 000 | 1 100 | 1 100 |
| 过程蒸汽产量/(kg/h) | 0 | 5 600 | 4 070 | 9 000 | 0 | 4 600 | 3 700 | 7 200 |
| 喷注蒸汽/(kg/h) | 4 320 | 4 320 | 0 | 0 | — | — | — | — |
| 燃气轮机燃料/(m$^3$/h) | 583 | 583 | 491 | 491 | 524 | 524 | 433 | 433 |
| 补燃燃料/(m$^3$/h) | 0 | 386 | 0 | 293 | 0 | 322 | 0 | 232 |
| 发电效率/% | 26.0 | 26.0 | 17.4 | 17.4 | 33 | 33 | 22 | 22 |
| 总效率/% | 26.0 | 54.4 | 73.0 | 87.9 | 57 | 37 | 79 | 86 |

# STIG 的新进展[*]

闻雪友

（哈尔滨船舶锅炉涡轮机研究所）

**摘　要**：燃气轮机回注蒸汽是一项有吸引力的发电技术，文中简要地介绍了这一领域内的最新总体进展情况。读者可从该技术在世界上现实发展这一角度来审视我国发展该项技术的必要性。

## 一、引言

在燃气轮机中回注蒸汽并不是全新的思想，最初是在地面燃气轮机上用于控制排放物中的 $NO_x$ 的含量，后来又发展到用于增大功率和提高效率。美国国际动力技术公司程大猷先生于 1976 年提出《双工质平行复合循环热机》专利发明（又称程氏循环）后，该技术真正引起商业兴趣。1985 年 1 月，国际上第一套回注蒸汽的商用机组正式投入商业运行，短短几年内，STIG 装置获得了相当迅速的发展。

目前，燃气轮机装置在热电联供和中心电站中应用较广。然而，简单循环燃气轮机因其部分负荷效率低，限制了将其更广泛地用作基本负荷机组，相对较低的纯发电效率也限制了将其更广泛地用作尖峰中心电站。STIG 技术在很大程度上克服了这些缺点，其高效率、热-电匹配的灵活性和投资成本低等优点使其在热电联供和中心电站供电方面具有很强的竞争力。

## 二、1985—1989 年间典型的 STIG 装置

国际上第一套双工质商用机组于 1984 年 11 月在美国加州大学安装验收完毕，1985 年 1 月该程氏循环热电联供装置投入商业运行。其由一台 Allison 501-KH 燃气轮机和余热锅炉组成。Allison 501-KH 系由 Allison 501-KB5 按喷注蒸汽运行的要求修改而成，喷注用的过热蒸汽用两根管子分别接到燃烧室的前后两排环形注汽管上，注入环管燃烧室，注汽量约 9 000 kg/h。在采用 STIG 装置后，功率由原先的 3 400 kW 增加到 5 400 kW，热效率

---

[*] 文章发表时间：1990 年 3 月。

由原来的 27% 提高到 37%。

1985 年 8 月,美国通用电气等五家公司在美国辛普森造纸厂的 LM5000 燃气轮机热电联供装置上成功地进行了喷注蒸汽试验,改进了燃气轮机的功率和效率。在此次改进经验基础上,1986 年 12 月,首台喷注蒸汽的 LM5000 STIG 燃气轮机安装在美国格罗国际蛋白技术公司的联合发电厂内。LM5000 STIG 的蒸汽注入方式有其独特的特点,分高压蒸汽系统和低压蒸汽系统。高压蒸汽系统中约有相当于 4% 空气流量的蒸汽从燃料喷嘴注入燃烧室主燃区,这对抑制排放物中的 $NO_x$ 含量有很好的作用。此外,约有相当于 3% 空气流量的蒸汽从压气机出口的扩散机匣处注入。在低压蒸汽系统中约有相当于 5% 空气量的低压蒸汽从低压涡轮导向叶片中注入,通过叶型叶盆上的排气孔进入燃气主流,类似于通常的空冷叶片。

LM5000 STIG80 的功率由 LM5000 的 332 101 kW 增加到 46 890 kW,热耗由 9 896 kJ/(kW·h)(9 380 Btu/(kW·h))降至 8 624 kJ/(kW·h)(8 174 Btu/(kW·h))。LM5000 STIG 120 功率已达 51 620 kW,热耗降至 8 342 kJ/(kW·h)(7 907 Btu/(kW·h))。

美国 GE 公司也已将 LM2500 改成回注蒸汽的 LM2500 STIG40,其功率由原来的 21 450 kW 增加到 27 013 kW,热耗由 9 981 kJ/(kW·h)(9 460 Btu/(kW·h))降至 8 754 kJ/(kW·h)(8 297 Btu/(kW·h))。

日本川崎重工株式会社把功率为 1 000 kW 的工业小燃气轮机 M1A-01 改为 M1A-01-CC 程氏循环后,功率增加到 1 750 kW,发电效率由 18% 增加到 26%。该装置已于 1988 年 4 月运行,其具有一单管回流式燃烧室。研究人员对注汽喷嘴三种不同的注汽方向排放物($NO_x$,CO,HC)以及燃烧室熄火特性的影响进行了试验。现在,川崎正在开发以 M1A-13 型(系 M1A-01 的发展)为基础的 M1A-13CC 程氏循环系统,功率将从 1 240 kW 增加到 2 370 kW,发电效率预计从 22% 提高到 33%。

近年来,英国 Ruston 公司在 TB5000 和 Tornado 发动机上也已提供了注蒸汽的选择。Ruston 公司根据注蒸汽的目的决定在发动机上装设什么系统。如仅需降低 $NO_x$,并不需要额外功率,则仅装设一次注汽系统,在燃烧室头部以汽、燃料约 1:1 的比例注入。如主要为提高功率,附带地进行 $NO_x$ 的控制,则在主燃区下游的掺混区中汽、燃料以大约 3:1 的比例进行二次喷注。如提高功率和控制 $NO_x$ 两者皆需,则同时安装两个喷注系统。

Tornado 注蒸汽后功率从 6 190 kW 提高到 7 200 kW,效率从 31.7% 提高到 32.2%。其第一套装置于 1988 年在日本某炼油厂运行。TB5000 注蒸汽后功率从 3 660 kW 提高到 4 470 kW,效率从 25.5% 提高到 26%。Ruston 公司选择的注汽量很小,因而距相应于最佳效率的汽气比较远,故效率增长甚微。

Stewart & Stevenson 公司正在进行 LM1600 STIG 发电机组的工作。TG1600-STIG50 型中余热锅炉产生 22.7 t/h 喷注蒸汽,ISO 条件下出力为 16 900 kW,热耗 8 938 kJ/(kW·h)(8 472 Btu/(kW·h)),预计 1990 年可供 60 Hz 发电机组。

## 三、趋势——ISTIG

在燃气轮机采用回注蒸汽技术的同时,再在低、高压压气机间采用间冷,这就是 ISTIG 装置。由于在压气机间设有中间冷却器,因而压气机的抽气温度要比相应简单循环时低,再加上蒸汽有较高的载热能力,故汽-气混合物作冷却介质时允许燃气初温提高而叶片金属温度仍维持原有水平,这就使 ISTIG 具有更高的效率。以 LM5000 为例,简单循环对其燃气初温约为 1 205 ℃,在 ISTIG 装置中燃气初温可升到 1 355 ℃,据 GE 公司的分析,LM5000 ISTIG 的效率将达 47%~48%,出力达 110 MW。表 1 中所列出的 LM5000 简单循环、LM5000 燃-蒸联合循环、LM5000 STIG120 循环及 LM5000 ISTIG 循环的对比会给人留下深刻的印象。

表 1  LM5000 系列燃气轮机性能对比

| 性能参数 | LM5000 简单循环 | LM5000 燃-蒸联合循环 | LM5000 STIG120 | LM5000 ISTIG |
|---|---|---|---|---|
| 功率/kW | 33 210 | 50 300 | 51 620 | 约 110 000 |
| 发电效率/% | 36.4 | 41.3 | 43.2 | 47~48 |

GE 公司根据对 ISTIG 概念的估计,预期经 3~4 年的发展可建成 LM5000 ISTIG 机组。GE 公司也正在考察 LM1600 ISTIG 的设计,在 ISO 条件下其连续发电功率达 40 000 kW(简单循环时为 13 400 kW),热效率接近 50%。

## 四、与煤电站、核电站、联合循环电站的比较

美国对国内采用 ISTIG 经济性分析的报告表明,以天然气为燃料的 ISTIG 装置能与新的燃煤电站、核电站相竞争,甚至天然气价格有明显增加时亦如此。煤电站、核电站较低的燃料价格被其明显高的投资成本和较低的发电效率所抵偿。

此外,ISTIG 装置的安装周期短、容量较小、高的可用率使其在用作公用电力事业方面具有灵活性。

在燃-蒸联合循环方面,现在的商用机组较好的是 250 MW,效率 40%,投资成本 500 美元/千瓦。正在建造的联合循环电站,燃气初温 1 200 ℃,效率 45%,投资成本 500 美元/千瓦。与 110 MW 的 ISTIG 相比,投资成本接近,但效率要低 2%~3%。

表 2 对中心电站各种发电技术做了综合比较。

表 2  中心电站发电技术比较

| 中心电站 | 功率/MW | 装置投资成本/(美元/千瓦) | 全负荷效率/% | 建设周期/年 | 可用率/% |
|---|---|---|---|---|---|
| 核电站 | 1 200 | 2 000 | 0.32 | 11 | 61.8 |
| 火电站 | 1 200 | 1 200 | 0.33 | 8 | 69.5 |
| 尖峰燃气轮机电站 | 150 | 300 | 0.29 | 3 | 80.6 |
| 联合循环电站 | 250 | 500 | 0.41 | 4 | 81.5 |
| 先进的联合循环电站 | 250 | 500 | 0.45 | 4 | 81.5 |
| 采用 ISTIG 装置电站 | 110 | 400~500 | 0.48 | 3 | 89.2 |

目前正在研究的最先进的联合循环发电装置可能是日本的一项国家计划：发展一型 122 MW 的带间冷、再热，有两个压气机、三个涡轮的复杂循环的燃气轮机，设计效率 47% ~ 48%，与 ISTIG 在同一范围内。计划是联合五台机的余热锅炉蒸汽通往一个大的蒸汽轮机，以 1 000 MW 的联合循环功率达到 50% 的效率，超过了第一代 ISTIG 的效率水平，然而这是以更高的投资、更大的复杂性和更大的尺寸为代价的。

## 五、发展前景

在世界范围内，认为 STIG 装置无论在热电联供或中心电站的应用方面均有重大的潜在市场。

一个较明显的事实是，天然气已在近 50 个发展中国家应用（包括 30 个石油进口国），这些国家中天然气的价格在 0.19 ~ 1.33 美元（0.20 ~ 1.44 美元/百万英热单位[①]），远低于美国 1984 年的平均价 2.37 美元/千克（2.50 美元/百万英热单位）。在这样低的天然气价格下，ISTIG 装置甚至可与电价最低的水电竞争。例如，巴西东南部一个新的水电装置的电价是 3 美分/（千瓦时），只要天然气的价格低于 2.18 美元/千克（2.30 美元/百万英热单位），ISTIG 装置就可与之竞争。

一个新的动向是 IGI - STIG 装置，其为一种把煤气化器与带间冷的回注蒸汽的燃气轮机相结合的概念。煤气发生器可从燃气轮压气机中的抽气来维持其气化过程，从余热锅炉中得到的蒸汽则维持其注汽过程。目前认为这一概念在技术上尚无不可克服之问题，只是花费甚大，估计要 1.25 亿 ~ 1.30 亿美元。GE 公司已完成 IGI - STIG LM5000 发电装置的初步设计评估。

STIG 装置的缺点是处理水的消耗问题，尽管对燃料节省与耗费处理水间的经济性分析以及 STIG 装置与其他类型装置水耗的比较早已有定论，且结论是明确肯定的，但在这方面的研究工作仍在进行。一个例子是带自除盐水的 STIG 装置的研究，除盐水借多效蒸馏装置或多级闪蒸装置产生，而这些装置的外部热源则是利用排气余能。

ARC（航空航天研究公司）近年来已在 Allison501 - KG 上进行了直接烧锯末的运行试验，并计划用注蒸汽来提高出力和效率。在投资成本低的燃气轮机上用低成本的生物燃料，再加上能大幅度提高功率的回注蒸汽技术，这种三合一的组合无疑是这一领域中重要的发展方向。

在我国，已有多个厂、所、院、校在 STIG 方面进行了多年的研究工作。最近哈尔滨船舶锅炉涡轮机研究所和哈船院联合，在 S1A - 02 工业小燃气轮机上完成了程氏循环改造，在国内首次实现了双工质平行 - 复合循环的整机运行和实验研究。

STIG 已成为一项突出的重要技术。

---

[①] 百万英热单位，即 MBtu，1 MBtu = $1\ 055 \times 10^6$ J。

# 燃气轮机余热锅炉市场趋向*

闻雪友

(哈尔滨船舶锅炉涡轮机研究所)

**摘　要**：通过对国外资料的整理、分析，向读者展示了世界燃气轮机余热锅炉市场的近况、概貌，同时也反映了燃气轮机余热锅炉的技术趋向。

## 一、引言

世界燃气轮机市场情况乐观。到1990年底，预计燃气轮机年订货量可达14 000 MW，仅燃气轮机的销售额就将超过40亿美元。大部分的订单是发电机组，包括简单循环和联合循环，机组供基本负荷或调峰机组使用。

由于销售的燃气轮机大部分用于热电联供或联合循环发电，因此与其配套的设备(如蒸汽轮机、余热锅炉、透平压缩机)还有约30亿美元的销售额，其中的余热锅炉绝大多数用于联合循环或工业热电联供装置。

文中所做的统计分析主要根据是1988年1月到1989年6月间订货及安装的226台余热锅炉资料。

## 二、趋向

过去，从燃气轮机排气中回收能量一直采取一种最少耗费的方式，最重要的准则是使初投资最低。然而，随着燃料价格增高，重点已转移到最大限度地回收能量上。这导致热回收技术的改进和发展，系统的研究变得更复杂，单压锅炉被更能有效地回收热量的多压锅炉设计所取代。由此，合乎逻辑的结果是：余热锅炉更加成为整个总系统设计的一个有机组成部分，而不是像以往那样，余热锅炉往往作为一种被选择的"尾端设备"。

---

\* 文章发表时间：1990年11月。

现在,余热锅炉甚至会影响到对燃气轮机的选择。例如,为使整个系统在部分负荷时保持高效率,要求燃气轮机的压气机采用进口可调导叶,以使部分负荷时余热锅炉进口的烟气温度基本保持不变。

与燃气轮机制造商在世界舞台上仅有为数不多的主角不同,锅炉制造商则是大量的,且没有很明显的世界市场领导。

过去二十年间,世界公用电力事业锅炉市场已有了重要变化,除了煤和核燃料锅炉外,转向现今的联合循环和流化床锅炉。常规蒸汽轮机电站用的大容量锅炉,现在已被联合循环的多压余热回收锅炉挤占了部分市场。

## 三、燃气轮机与余热锅炉的配套

表1列出了226套设备中订货量在6台套以上的燃气轮机型号。应该指出,即使是同一型号的燃气轮机,根据用途的不同,与其相配的余热锅炉也可能不同。此外,该数据也不代表该型燃气轮机在该期间内的总销售数,仅指带余热锅炉的特定应用场合。

表1 燃气轮机与余热锅炉配套情况

| 燃气轮机型号 | ISO 额定功率/kW | 配套台份 |
| --- | --- | --- |
| MS6001 | 38 340 | 30 |
| LM2500 | 21 700 | 19 |
| V94 | 150 200 | 15 |
| TYPE11N | 81 600 | 11 |
| CENTAURH | 3 880 | 13 |
| TORNADO | 6 249 | 13 |
| MARS | 8 840 | 10 |
| MS5001 | 26 830 | 10 |
| MS9001E | 116 900 | 10 |
| LM5000 | 33 762 | 10 |
| W501 D5 | 104 400 | 8 |
| SB5 | 1 140 | 7 |
| CX350－KB5 | 3 731 | 6 |
| CW251 | 42 500 | 6 |
| MS7001 | 83 500 | 6 |

## 四、自然循环与强制循环

燃气轮机的余热锅炉究竟采用自然循环方式还是强制循环方式,各客户和制造商有不同的选择和理由。一般来说,美国的制造商大部分选择水平布置的自然循环设计,而欧洲的供应商喜欢采用垂直布置的强制循环。对这种传统的一种解释是:历史地看,美国的公用电力事业在现场选择方面一直比其欧洲、日本竞争对手有更多的自由,更大的现场空间促使锅炉的户外设计采用水平布置和自然循环。欧洲和日本的公用电力事业面对一个较严峻的现场条件,因而采用更紧凑的垂直布置,并且把大型燃煤蒸汽轮机电站所用的传统的强制循环移到余热锅炉中来。

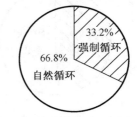

**图 1** 余热锅炉循环方式的统计

图 1 表示 226 套装置中水平布置自然循环与垂直布置强制循环锅炉各自所占的百分比。

## 五、单压与多压余热锅炉

采用多压余热锅炉设计技术可更有效地回收燃气轮机的排气能量,当然也意味着投资成本增加。

能量平衡分析表明,对燃气轮机排气温度在 530~580 ℃ 范围内,用双压系统来取代单压系统,可使传热能量损失大致从 15% 的排气可用能率降为 8%,相应于热力学第一定律的锅炉效率约提高 3 个百分点。采用三压来进一步回收是可能的,但其增益量减小。在上述排气温度范围内,相对于一个已最佳化的双压系统而言,采用三压系统后,其第一定律的效率增益约为 1 个百分点。

图 2 为 226 台余热锅炉中单压和多压系统各自所占的百分比,两者几乎是平分秋色。

图 3 为多压余热锅炉中双压与三压各自所占的百分比,可见双压应用得更为广泛,但是三压也已并不少见。

表 2 列出了 62 台用于燃-蒸联合循环的余热锅炉设计情况,采用单压设计的几乎为零。

**表 2 联合循环中余热锅炉形式统计**

| 余热锅炉形式 | 单压 | 双压 | 三压 |
|---|---|---|---|
| 数 量 | 1 | 49 | 12 |

图 4 表示 226 台余热锅炉中无补燃的和带补燃余热锅炉各自所占的百分数。根据热电联供的准则,以及独立的电力和工艺流程设备的要求,当余热回收系统仅靠燃气轮机排气已不能产生所需的蒸汽时,可采用补燃方式来满足。

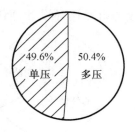 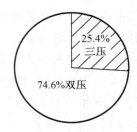 

图 2　余热锅炉单、多压系统的统计　　图 3　多压余热锅炉系统中双压、三压的统计　　图 4　无补燃、带补燃余热锅炉的统计

## 六、分布

统计的 226 台装置的安装现场地区分布如图 5 所示。很显然,在天然气资源丰富的地区,燃气轮机热电联供或联合循环得到了广泛应用。图 6 表示应用的分类情况,工业热电联供最为普遍。

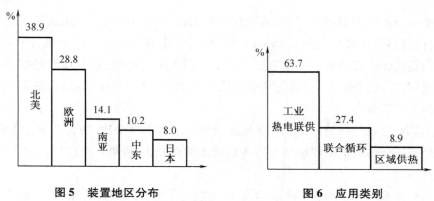

图 5　装置地区分布　　　　　　　图 6　应用类别

## 七、结束语

用户根据自己的需要进行工程选择时,常常希望了解这些类似装置的工程实际应用情况。对于燃气轮机的余热锅炉,讨论的问题包括自然循环或强制循环,用单压或多压系统,是否采用补燃等。文中的统计分析以一种直截了当的方式提供了这方面的参考。

统计资料虽不完全,但是已明显地反映了燃气轮机余热锅炉的技术趋向。

# 涡轮转子整圈带冠叶片振动分析*

闻雪友  刘 岩
（哈尔滨船舶锅炉涡轮机研究所）

**摘 要**：讨论变截面扭转带冠叶片整圈振动问题。在 Hall R M 工作的基础上，提出当量叶片的方法，从而得到一种可在计算机上快速计算变截面扭转整圈带冠叶片振动频率的简便估算方法。

## 一、引言

带冠叶片已在燃气轮机、蒸汽轮机上广泛应用。通过叶冠间的相互作用使叶片联成整圈，无疑在改善叶片的振动强度方面具有明显的优点，但其振动特性较复杂。

Hall R M[1]提出了一种计算双对称等截面叶片带 Z 形冠（忽略冠的质量）的整圈叶片振动的方法，但显然尚不够实用。文中在此基础上针对在实际的涡轮机械上应用最广泛的 Z 形冠和平行四边形冠，以及叶冠间可能存在间隙、松接（冠间无分离，但沿接触面可发生滑动）和紧接（冠间接触面上无相对滑动）及冠间的线性摩擦等边界条件，对等截面带冠直叶片、带冠扭叶片（自然扭转率较小的）并计入冠的质量影响的实际条件下进行整圈叶片的振动分析。其基本思路是引入当量叶片的方法做近似处理，加上闭合的周期性结构的特点，使计算工作甚为简便。

文中对一个实际的具有变截面带冠扭叶片的涡轮转子进行了测频试验，并对试验结果与相应的计算数据进行了比较。

符号说明如下。

| | |
|---|---|
| $A$ | 当量叶片横截面积。 |
| $EI_y$、$EI_z$ | 叶片各自在 $y,z$ 方向的弯曲刚度。 |
| $F_{yn}$、$F_{zn}$ | 作用在第 $n$ 个冠 $y$ 和 $z$ 方向上叶片的力。 |
| $J_c$ | 当量叶片单位长度截面的极惯性矩。 |

---

\* 文章发表时间：1993 年 2 月。

| | |
|---|---|
| $GJ_p$ | 当量叶片单位长度翼型扭转刚度。 |
| $P_n$ | 在左端第 $n$ 个冠接触面平行于接触面的力。 |
| $Q_n$ | 在右端第 $n$ 个冠接触面正交于接触面的力。 |
| $T_n$ | 作用在第 $n$ 个叶片上的转矩。 |
| $d$ | 从叶片扭转中心到接触面的距离。 |
| $y_n$ | 在 $y$ 方向第 $n$ 个冠的位移。 |

$S = \cos p_y + \cosh p_y$

$\bar{S} = \cos p_z + \cosh p_z$

$u = \cosh p_y - \cos p_y$

$\bar{u} = \cosh p_z - \cos p_z$

$v = \sin p_y - \sin p_y$

$\bar{v} = \sinh p_z - \sin p_z$

$w = \sinh p_y + \sin p_y$

$\bar{w} = \sinh p_z + \sin p_z$

$a = [(EI_y/GJ_p) \cdot (J_c/AL^2)]^{1/2}$

$J = (-1)^{1/2}$

| | |
|---|---|
| $K$ | 节点直径数。 |
| $L$ | 叶片长。 |
| $m$ | 叶片数。 |
| $f$ | 自然频率。 |
| $t$ | 时间。 |
| $\alpha$ | 叶冠接触面角度。 |
| $\gamma$ | 当量叶片安装角。 |
| $\mu$ | 质量密度。 |

$\varepsilon = \alpha - \gamma$

$p_y = (\mu L^4 P^2 A/EI_y)^{1/4}$

$p_z = p_y(I_y/I_z)^{1/4}$

| | |
|---|---|
| $\theta_n$ | 第 $n$ 个冠的转角。 |

$\varphi = 2K\pi/m$

| | |
|---|---|
| $f_1$ | 单叶片精确计算值。 |
| $f_2$ | 以当量截面几何特性作等截面叶片的计算频率值。 |
| $I_{min}$ | 当量叶片的最小惯性矩。 |
| $I_r$ | 当量叶片的极惯矩。 |
| $I_{r_p}$ | 当量叶片弯心的极惯矩。 |
| $\beta_顶$ | 实际叶片顶部安装角。 |
| $\beta_根$ | 实际叶片的根安装角。 |

$z_n$　　　　　　在 $z$ 方向第 $n$ 个冠的位移。
$\mu_0$　　　　　　叶冠接触面间的线性摩擦系数。

## 二、当量叶片法

(一) Hall R M 工作的基本要点

Hall R M 对叶片做了如下假定：

(1) 叶片是无扭转双对称等直截面；
(2) 忽略叶冠的质量；
(3) 沿垂直叶冠接触面和接触面方向的弯矩为零；
(4) 叶片被固定在刚度为无限大的盘上；
(5) 相邻叶冠接触面是不分离的，即载荷充分大；
(6) 接触面的长度比接触面的中心到叶片的扭转中心的距离小很多，目的是避免接触面的斜向及倾斜的综合作用；
(7) 接触面的中心在相邻两个叶片的扭转中心连线上。

Hall R M 考虑了两种极端的情况：

(1) 叶冠的接触面紧连接，即铰支，如图 1 所示；
(2) 相对无摩擦，自由滑动，即松连接，如图 2 所示。

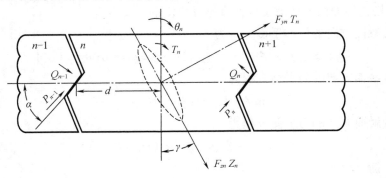

**图 1　紧连接**

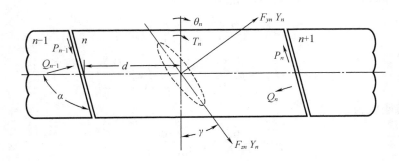

**图 2　松连接**

## （二）当量叶片法的基本思路

计算非对称变截面扭曲叶片"当量叶片"方法的要点如下。

（1）叶冠质量的考虑：用一单叶片的计算频率第一音调作为第一阶零节径的修正值。

（2）扭转中心仍假定在叶冠中心。

（3）对非对称变截面叶型的处理是寻找一当量叶片，其振动与单只叶片的第一音调振动非常接近。

当量叶片的安装角由下式确定：

$$\gamma = \frac{1}{3}\beta_{顶} + \frac{2}{3}\beta_{根}$$

当量叶片的几何特性则由以下公式确定：

$$A = R^2 A_0$$
$$I_{\min} = R^4 I_{\min}$$
$$I_r = R^4 I_{r_0}$$
$$I_{r_p} = R^4 I_{r_{y0}}$$

脚码"0"系 $\gamma$ 角所在截面的几何特性，$R$ 值则依据频率相等原则确定，即

$$\frac{f_1}{f_2} = 1$$

（4）对于松连接的情况考虑了线性摩擦导出新的振动方程。

垂直于接触面方向的位移平衡方程为

$$(z_n - z_{n+1})\cos\xi + (y_n - y_{n+1})\sin\xi = d\cos\alpha(\theta_{n+1} + \theta_n) \tag{1}$$

垂直于接触面方向的力平衡方程为

$$F_{y,n}\sin\xi + F_{z,n}\cos\xi = (Q_n - Q_{n-1}) \tag{2}$$

平行于接触面方向的力平衡方程为

$$F_{y,n}\cos\xi - F_{z,n}\sin\xi = \mu_0(Q_n - Q_{n-1}) \tag{3}$$

扭矩的平衡方程为

$$T_n = d(Q_n + Q_{n-1})\cos\alpha - d\sin\alpha\mu_0(Q_n + Q_{n-1}) \tag{4}$$

由古典薄梁理论知，在 $y, z$ 方向动刚度可表示为

$$F_{y,n} = p_y^3 \frac{EI_y}{L^3} \frac{(s^2 - wv)}{(vs - wu)} y_n$$

$$F_{z,n} = p_z^3 \frac{EI_z}{L^3} \frac{(\bar{s}^2 - \bar{w}\bar{v})}{(\bar{v}\bar{s} - \bar{w}\bar{u})} z_n \tag{5}$$

叶顶扭转刚度为

$$T_n = \frac{GJ_p}{L} a p_y^2 \frac{\cos(\alpha p_y^2)}{\sin(\alpha p_y^2)} \tag{6}$$

由式(3)~式(5)得

$$\frac{z_n}{y_n} = \left(\frac{p_y}{p_z}\right)^3 \left(\frac{I_y}{I_z}\right) \frac{(s^2-wv)}{(vs-wu)} \cdot \frac{(\bar{v}\bar{s}-\bar{w}\bar{u})}{(\bar{s}^2-\bar{w}\bar{v})} \frac{(\cos\xi-\mu_0\sin\xi)}{(\sin\xi+\mu_0\cos\xi)} \tag{7}$$

将式(5)、式(6)分别代入式(2)、式(4)整理得

$$(\theta_n+\theta_{n-1})\frac{GJ_p ap_y^2}{d(\cos\alpha-\mu_0\sin\alpha)}\frac{\cos(ap_y^2)}{\sin(ap_y^2)} +$$

$$p_y^2 EI_y \frac{(s^2-wv)}{(vs-wu)} \cdot \left[(\sin\xi+\mu_0\cos\xi) + \frac{(\cos\xi-\mu_0\sin\xi)\cos\xi}{\sin\xi}\right](y_n+y_{n-1})$$

$$= 2(Q_n+Q_{n-1}) \tag{8}$$

由式(6)、式(4)得

$$(\theta_n-\theta_{n-1}) = \frac{EI_y}{GJ_p}\frac{p_y}{L^2 a}\frac{d(\cos\alpha-\mu_0\sin\alpha)}{\sin\xi} \cdot \frac{(s^2-wv)}{(vs-wu)} \frac{\sin(\alpha p_y^2)}{\cos(\alpha p_y^2)} \cdot (y_{n+1}+2y_n+y_{n-1}) \tag{9}$$

由式(1)得

$$(z_{n+1}-2z_n+z_{n-1})\cos\xi + (y_{n+1}-2y_n+y_{n-1})\sin\xi + (\theta_{n+1}-\theta_{n-1})d\cos\alpha = 0 \tag{10}$$

由式(8)~式(10)整理得

$$y_{n+1} - 2y_n\frac{(D-H)}{(D+H)} + y_{n-1} = 0 \tag{11}$$

式中

$$D = \left(\frac{p_y}{p_z}\right)^3 \left(\frac{I_y}{I_z}\right)\frac{(s^2-wv)}{(sv-wu)}\cos\xi(\cos\xi-\mu_0\sin\xi) + \frac{(\bar{s}^2-\bar{w}\bar{v})}{(\bar{s}\bar{v}-\bar{w}\bar{u})}\sin\xi(\sin\xi+\mu_0\cos\xi)$$

$$H = \left(\frac{p_y}{a}\right)^3 \left(\frac{EI_y}{GJ_p}\right)\frac{d\cos\alpha(\cos\alpha-\mu_0\sin\alpha)}{L} \cdot \frac{\sin(\alpha p_y^2)}{\cos(\alpha p_y^2)} \frac{(s^2-wv)}{(vs-wu)} \frac{(\bar{s}^2-\bar{w}\bar{v})}{\bar{v}\bar{s}-\bar{w}\bar{u}} \tag{12}$$

利用行波解设

$$y_n = \hat{y}e^{jft},\ y_{n+1} = \hat{y}e^{j(ft+\varphi)},\ y_{n-1} = \hat{y}e^{j(ft-\varphi)} \tag{13}$$

或

$$y_{n+1} = y_n e^{j\varphi},\ y_{n-1} = y_n e^{-j\varphi}$$

将式(12)、式(13)代入式(11)整理得

$$(1-\cos\varphi)\left[(sv-wu)(\bar{s}^2-\bar{w}\bar{u})\sin\xi(\sin\xi+\mu_0\cos\xi) + \left(\frac{p_y}{p_z}\right)^3\left(\frac{I_y}{I_z}\right)(s^2-wu)(\bar{v}\bar{s}-\bar{w}\bar{u}) \cdot \right.$$

$$\left. \cos\xi(\cos\xi-\mu_0\sin\xi)\right]\cos(\alpha p_y^2) - (1+\cos\varphi)\left(\frac{EI_y}{GJ_p}\right)\left(\frac{p_y}{a}\right)\frac{d\cos\alpha(\cos\alpha-\mu_0\sin\alpha)}{L} \cdot$$

$$(s^2-wv)(\bar{s}^2-\bar{w}\bar{v})\sin(\alpha p_y^2) = 0 \tag{14}$$

同理,得到紧连接情况下的振动方程为

$$(1-\cos\varphi)\left[(sv-wu)(\bar{s}^2-\bar{w}\bar{u})\sin^2\xi + \left(\frac{p_y}{p_z}\right)^3\left(\frac{I_y}{I_z}\right)(s^2-wu)(\bar{v}\bar{s}-\bar{w}\bar{u})\cos^2\xi\right] \cdot$$

$$\cos(\alpha p_y^2) - (1+\cos\varphi)\left(\frac{EI_y}{GJ_p}\right)\left(\frac{P_y}{a}\right)\left(\frac{d\cos\alpha}{L}\right)^2(s^2-wv)(\bar{s}^2-\bar{w}\bar{v})\sin(\alpha p_y^2) = 0 \tag{15}$$

## 三、算例

某发动机的动力涡轮转子是一个单盘转子,盘装有 112 片叶片,叶片与轮盘由纵树型叶根相连接。在对叶轮刚性固定的条件下测定了轮系切向带节径的弯曲振动(A 型振动)及 B 型振动。

摩擦系数取为 $\mu_0$ 取 0.45,0.30。

计算结果及其与实测值的比较如表 1、表 2。

**表 1　A 型振动情况下结果比较**

| 频率值/Hz | | 节径数 | | | | | | |
|---|---|---|---|---|---|---|---|---|
| | | 0 | 1 | 2 | 3 | 4 | 5 | 6 |
| A 型频率 | 实验值 | 97 | 113 | 127 | 139 | 150 | 160 | 170 |
| | 无摩擦当量法 | 92 | 111 | 151 | 192 | 229 | 258 | 282 |
| | 有摩擦当量法 $\mu_0 = 0.45$ | 90 | 97 | 115 | 139 | 164 | 190 | 212 |
| | 有摩擦当量法 $\mu_0 = 0.30$ | 92 | 102 | 124 | 152 | 181 | 208 | 232 |

**表 2　B 型振动情况下结果比较**

| 频率值/Hz | | 节径数 | | | | | | |
|---|---|---|---|---|---|---|---|---|
| | | 0 | 1 | 2 | 3 | 4 | 5 | 6 | ⋯ |
| B 型频率 | 计算值 | 334 | 334 | 335 | 336 | 337 | 338 | 340 | ⋯ |
| | 实验值 | 334 | ⋯ | ⋯ | | | | | 370 |

## 四、计算结果分析

(1) B 型振动计算结果与实验结果比较一致。

(2) 叶片的当量角 $\gamma$ 对松接情况下频率影响是很大的。

(3) 如不考虑摩擦,则计算值同实验值有比较大的差别;如考虑摩擦,则计算值有明显的改善,但总的趋向是误差随着节径数的增加而增大(图 3)。

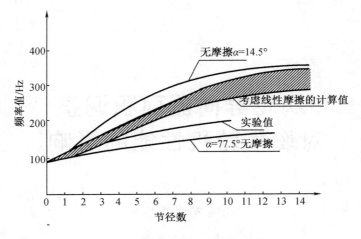

图3 有无摩擦时计算误差与节径数的关系

# 参 考 文 献

[1] HALL R M, ARMSTRONG E K. The vibration characteristics of an assembly of interlock shrouded turbine blades[J]. Structural dynamic aspects of bladed disk assemblies, 1976: 1 – 13.

# 涡轮导向器面积调整对燃气轮机性能的影响*

闻雪友　钱振岳

**摘　要**：讨论带动力涡轮的双轴燃气轮机中，涡轮导向器面积变化对燃气轮机性能的影响。介绍一种工程计算方法，这种方法的精度已在一台大功率燃气轮机上进行了试验并得到部分验证。

## 一、引言

在研制燃气轮机的过程中，保证发动机达到预定的功率、油耗等指标以及在整个运行范围内无喘地工作是一个基本的要求。事实上，由于设计、制造、装配等各环节中的种种因素，总可能导致发动机的主要数据与技术要求有某种程度的不符，因此需要一些简单、实用的调整手段。

利用改变涡轮导向器喉部面积可以达到上述目的。因为在一台已加工完毕的发动机上，导向器面积仍有可能较容易地改变，而且这种变化会对发动机的参数产生重要影响。因而，涡轮导向器面积调整这种方法在实践中广为采用，可以利用该方法向所希望的方向修正发动机的参数（功率、油耗、喘振裕度、燃气初温等），补偿其他参数（效率、压力损失等）变化的影响，调整匹配关系。为便于调整，许多发动机的涡轮导向器上已设计有各种类型的可供调整的结构。

涡轮导向器面积对通流部分各截面参数和发动机总性能的影响能力及其实施可调的现实性，使得可变几何涡轮在发动机上得到采用。根据循环线路，适当选择可变几何涡轮级的位置可获最佳增益。按其使用场合和要求，可使发动机具有改善部分负荷经济性，改善启动性、加速性，提高制动能力，提供动力涡轮超速保护等一系列优点。

需要一种工程计算方法，且该方法是较简单而又具有一定的精度，能预示导向器面积调整后发动机性能变化情况。采用将压气机特性和导向器面积变化后的涡轮特性相叠加的方法可

---

\* 文章发表时间：1981年12月。

以获得精确的解,但在对各级导向器面积做综合调配的多方案计算中会使求解有所不便,当缺乏确切的压气机特性线时则更显困难。文献[1]提出的工程小偏差计算法把发动机工作过程方程化为小偏差形式,把联系工作过程参数的复杂方程组的求解化为联系参数与其原始值偏差量的线性方程的求解,使计算工作量大为缩减。利用所得的影响系数表来分析问题则较为简明。因而文献[1]所提出的工程小偏差法在航空工业中得到了广泛应用。

文中讨论带动力涡轮的双轴燃气轮机中如何适当地运用小偏差计算法,并通过实验验证了其计算精度。

## 二、计算方法

小偏差法是使表示某种现象的关系式线性化的一种方法。因而,当我们把发动机的工作过程方程化为小偏差形式时,无论方程式或未知数的数目都没有改变,故此小偏差方程组的可解性与原方程组相同。

文献[1]详述了发动机工作过程的小偏差方程及涡轮逐到小偏差方程的导出,文中不再赘述。文中所讨论的带动力涡轮的双轴燃气轮机类似于航空双轴涡轮螺旋桨发动机。文献对后者是用如下方法来处理的:基于现代涡轮螺旋桨发动机燃气初温高,动力涡轮部分的膨胀比往往很高,表明动力涡轮第一级导向器的膨胀比和 $q^0(\lambda'_{c,a})$ 随透平总膨胀比的变化很少。因而发动机通流部分中,从压气机进口到动力涡轮第一级导向器的部分,可以看作一个带临界流动工况喷口的涡轮喷气发动机。此时,把动力涡轮一级导向器的喉部面积看作为喷管和面积,并认为此截面的 $q(\lambda)$ 值为常数,取动力涡轮进口压力与大气压力之比 $\pi_\Sigma$ 作为喷管出口截面内外压力比 $\pi_e$。然后再对动力涡轮建立方程,求解。

对上述计算方法做进一步的精化。因为许多双轴燃气轮机其动力涡轮一级导向器喉部截面处的流动状态并非非常接近临界,在任何情况下都把动力涡轮看作一个临界喷管影响了计算结果的精度,有时会带来不小的误差。在讨论导向器面积变化的影响时也最好计入 $q(\lambda_{c,a})$ 变化的影响,因为其随导向器面积的变动而有较显著的变化,部分地抵消了导向器变化的影响能力。

通过把工作过程小偏差方程与涡轮逐列小偏差方程相结合的方法可以实现这一点。当然,所需的方程数将成倍地增加。用这样的方法可对图1所示的双轴燃气轮机列出小偏差方程组。图中高压涡轮为两级,动力涡轮一级,为使方程组尽可能地包括一般情况,高、低压涡轮间尚有中间连接管(一些重型及航空发动机改装的燃气轮机具有这种形式)。

下面,列出在不变的燃气发生器折合转速下的小偏差方程组。符号含义见附录Ⅰ,方程组中各系数列于附录Ⅱ。

燃气发生器压气机、涡轮功率平衡方程为

$$K_3 \delta\pi_{T_1} + (K_{11} - K_1)\delta\pi_K + \delta T_\Gamma = -\delta\eta_K - \delta\eta_{T_1} - \delta q$$

压比方程为

$$\delta\pi_\Sigma - \delta\pi_K + \delta\pi_{T_1} = \delta\sigma_B + \delta\sigma_\Gamma + \delta\sigma_Д$$

高压涡轮Ⅰ级导向器与动力涡轮Ⅰ级导向器间的连续方程为

$$\left(1 - \frac{1}{2}K_3 K_4\right)\delta\pi_{T_1} - \delta q(\lambda'_{c,a}) + \delta q(\lambda_{c,a}) = \delta\sigma_\text{Д} + \delta F'_{c,a} - \delta F_{c,a} + \frac{1}{2}K_4 \delta\eta_{T_1}$$

压气机特性线方程为

$$\delta G_B - K_{10}\delta\pi_K = \delta\sigma_B$$

高压涡轮和压气机间的连续方程为

$$(1 - K_{10})\delta\pi_K - \frac{1}{2}\delta T_\Gamma + \delta q(\lambda_{c,a}) = \delta q - \delta\sigma_\Gamma - \delta F_{c,a}$$

压气机压缩过程空气温升方程为

$$\delta T_K + (K_{11} - K_1)K_2 \delta\pi_K = -K_2 \delta\eta_K$$

高压涡轮膨胀过程燃气温降方程为

$$\delta T_T - \delta T_\Gamma + K_3 K_4 \delta\pi_{T_1} = -K_4 \delta\eta_{T_1}$$

燃烧过程方程为

$$\delta G_T - \delta G_B - K_5 \delta T_\Gamma + (K_5 - 1)\delta T_K = 0$$

动力涡轮压比方程为

$$\delta\pi_C - \delta\pi_\Sigma + \delta\pi_{T_2} = \delta\sigma_C$$

动力涡轮Ⅰ级导向器与排气管出口截面间的连续方程为

$$\delta\pi_{T_2} - A_Z K'_6 \pi_\Sigma + A_Z q(\lambda'_{c,a}) = A_Z(1 + K'_6)\delta\sigma_C + \frac{1}{2}A_Z B_4 \delta\eta_{T_2} - A_Z \delta F'_{c,a}$$

动力涡轮功率方程为

$$\delta N_B - \delta G_B - \delta T_T - B_3 \delta\pi_{T_2} = \delta q + \delta\eta_{T_2} + \delta\eta_m$$

耗油率方程为

$$\delta C_e - \delta G_T + \delta N_B = 0$$

高压涡轮压比方程为

$$\delta\pi_{T_1} - \delta\pi_1 - \delta\pi_2 - \delta\pi_3 - \delta\pi_4 = 0$$

高压涡轮Ⅰ级导向器与Ⅰ级动叶间的连续方程为

$$(1 + b_1 - Z_1)\delta\pi_1 - b_2 \delta\pi_2 = \delta F_2 - \delta F_{c,a}$$

高压涡轮Ⅰ级动叶与Ⅱ级导向器间的连续方程为

$$(1 + b_2 - Z_2)\delta\pi_2 - b_3 \delta\pi_3 = \delta F_3 - \delta F_2$$

高压涡轮Ⅱ级导向器与Ⅱ级动叶间的连续方程为

$$(1 + b_3 - Z_3)\delta\pi_3 - b_4 \delta\pi_4 = \delta F_4 - \delta F_3$$

用压比表示的气动函数方程为

$$\delta q(\lambda_{c,a}) - b_1 \delta\pi_1 = 0$$

$$\delta q(\lambda_2) - b_2 \delta\pi_2 = 0$$

$$\delta q(\lambda_3) - b_3 \delta\pi_3 = 0$$

$$\delta q(\lambda_4) - b_4 \delta\pi_4 = 0$$

高压涡轮 I 级压比方程为
$$\delta\pi_{1T} - \delta\pi_1 - \delta\pi_2 = 0$$
高压涡轮 II 级压比方程为
$$\delta\pi_{2T} - \delta\pi_3 - \delta\pi_4 = 0$$
动力涡轮压比平衡方程为
$$\delta\pi_{T2} - \delta\pi_1' - \delta\pi_2' = 0$$
动力涡轮 I 级导向器与 I 级动叶间的连续方程为
$$(1 + b_1' - Z_1')\delta\pi_1' - b_2'\delta\pi_2' = \delta F_2' - \delta F_{c,a}'$$
用压比表示的气动函数方程为
$$\delta q(\lambda_{c,a}') - b_1'\delta\pi_1' = 0$$
$$\delta q(\lambda_2') - b_2'\delta\pi_2' = 0$$
动力涡轮燃气膨胀温降方程为
$$\delta T_N - \delta T_T + B_3 B_4 \delta\pi_{T2} = -B_4 \delta\eta_{T_2}$$

上述方程组中列于等号右边的均为独立变量。文中讨论导向器喉部面积变化的影响，所讨论的独立变量是涡轮导向器面积以及涡轮效率的变化(由导向器面积变动引起的)。

用元素按列存放列主元高斯消去解此线性方程组较为简便，所求得的结果即为影响系数，其表示在某一独立变量发生变化时(而其余的独立变量是不变的)所引起的随变量的变化倍数。

可以方便地对任意级数的涡轮按上述方法写出小偏差方程。

## 三、试验研究

在一台大功率燃气轮机上进行了动力涡轮导向器面积调整对发动机性能影响的试验研究。整个试验装置包括燃气轮机、减速器、测功器。试验装置的示意图如图 1 所示。双级的高压涡轮带动压气机，动力涡轮为单级，高压涡轮与动力涡轮间有一中间扩压器，动力涡轮转子经输出轴与减速器相连，减速器的输出端与水力测功器连接。

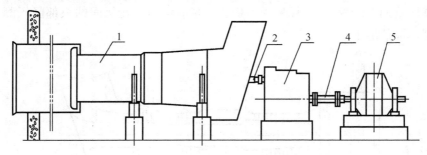

**图 1　试验装置示意图**

1—试验燃气轮机；2—动力涡轮输出轴；3—减速器；4—减速器输出轴；5—水力测功器

动力涡轮导向器面积的变动是用偏心衬套实现的(图 2)。导叶精铸而成，上下均带有缘板。上缘板上有两个螺孔,通过机上的孔拧入上缘板前螺孔的定位螺钉与插入内气封环上的下缘板短轴构成了旋转轴线。当后端孔上装入偏心衬套时,转动偏心衬套即可使偏心距分别处于两个极端位置,分别获得最大和最小面积。但限于结构,面积的最大改变量仅4%。

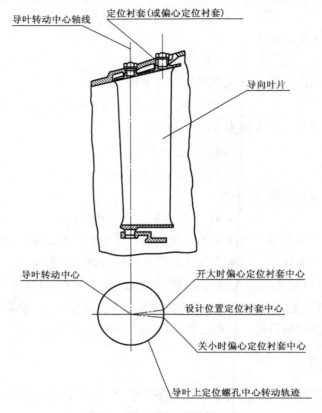

**图 2　导向器面积调整机构**

发动机的各测量截面及各测量截面上的测点布置如图 3 所示,图中各代号说明如表 1 所示。功率测量用电子秤,燃油耗量用容积法计量。动力涡轮级间的根部静压测压孔位于中间扩压器后隔热屏上。

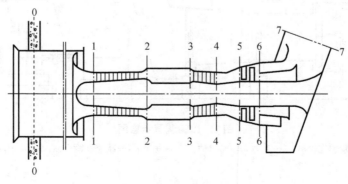

**图 3　测点布置图**

表1 图3说明

| 序号 | 截面 | 位置名称 | 外壁静压 | 总压点数 | 总温点数 | 备注 |
|---|---|---|---|---|---|---|
| 0 | 0—0 | 稳流室 | — | — | 7 | 1支0.1℃汞温度计6支铂电阻温度计 |
| 1 | 1—1 | 进口导叶前 | 4 | 2×5 | — | |
| 2 | 2—2 | 压气机出口 | 4 | 1×5 | 4×5 | |
| 3 | 3—3 | 高压涡轮进口 | — | — | — | |
| 4 | 4—4 | 高压涡轮输出口 | — | — | — | |
| 5 | 5—5 | 低压涡轮进口 | 4 | 4×5 | 4×7 | |
| 6 | 6—6 | 低压涡轮输出口 | 5 | 4×5 | 4×7 | |
| 7 | 7—7 | 排气管出口 | 4 | — | — | |

试验分别在对导向器处于额定位置、开大2%、关小2%三种状态下进行。在每种状态下，在35%～100%的功率范围内进行了测量，这时动力涡轮的转速以两种方式变化：一种是控制动力涡轮的转速按螺旋桨的功率——转速特性变化；另一种是在燃气发生器转速保持不变的情况下广泛变动动力涡轮的转速，以此获得燃气轮机的外特性。

由于导向器面积改变量较小的特点，为减少测量误差在整个测试过程中测试系统不做任何更改，发动机上除动力涡轮导向器面积变化外也无别的变动。

试验结果整理成图线，分别为动力涡轮导向器面积变化对流量、压比、涡轮膨胀比、燃气初温、发动机功率、耗油率、动力涡轮级向根部静压以及发动机外特性的影响曲线。

## 四、讨论

导向器喉部面积变化影响通流参数。当动力涡轮导向器喉部面积增大时流量增加（图4），压比则减小（图5）。这时由于高压涡轮的膨胀比增大（图6），为保持折合转速不变所需的燃气初温降低（图7），而动力涡轮的膨胀比又因导向器面积增大而减小（图6），最终导致的结果是功率下降、油耗增加（图8）。

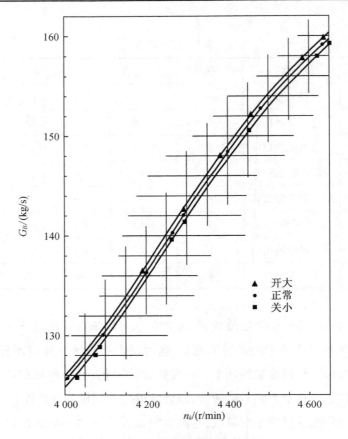

图 4　$F'_{c,a}$ 变化对空气流量的影响

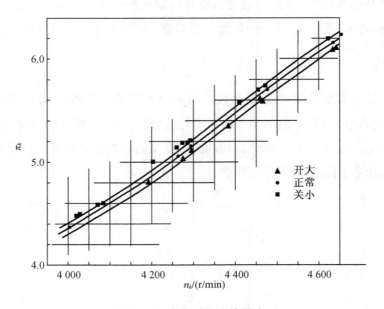

图 5　$F'_{c,a}$ 变化对压比的影响

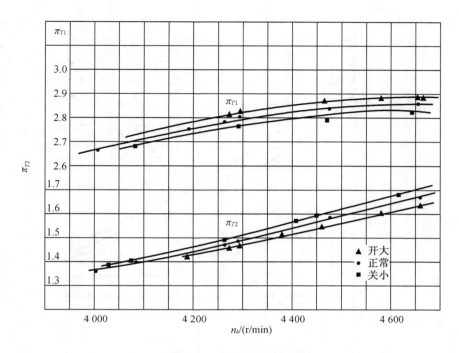

图 6  $F'_{c,a}$ 变化高压及动力涡轮膨胀比的影响

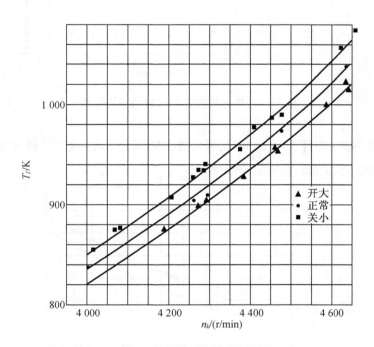

图 7  $F'_{c,a}$ 变化对燃气初温的影响

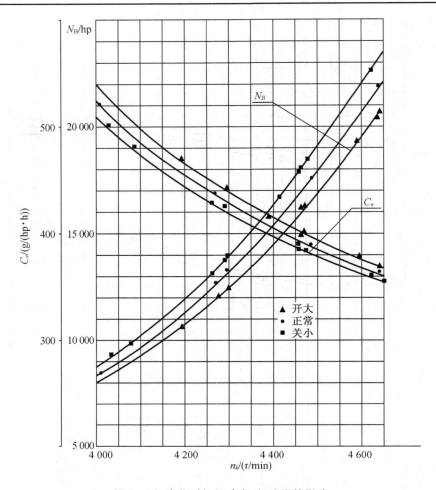

**图 8** $F'_{c,a}$ 变化对机组功率、耗油率的影响

在动力涡轮导向器面积增大的同时,导向器的膨胀比下降,级的反动度增加,因而所测得的动力涡轮级间根部静压值增加(图9)。

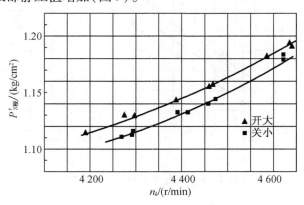

**图 9** $F'_{c,a}$ 变化对级间根部静压的影响

当动力涡轮导向器面积减小时则上述情况向相反方向变化。

在导叶转动的同时涡轮效率发生变化,由于试验中面积变化仅±2%,效率变化甚微(仅约0.1%),只能从动力涡轮特性图(图10)上来观察导叶转动前后效率网线的变动趋势。

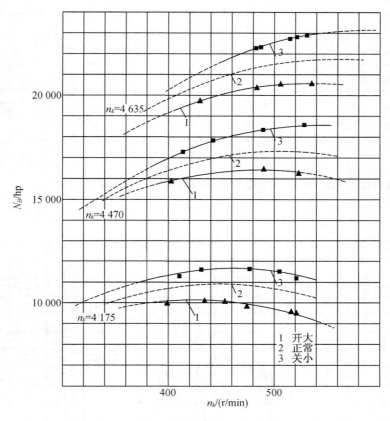

图10　$F'_{c,a}$变化时动力涡轮特性图的变化

根据试验结果整理的影响系数列于表2,表中还列入了作为试验装置的燃气轮机设计点的影响系数计算结果。作为对比还列入了把动力涡轮导向器看作临界喷口时的计算结果,可以看出,本计算结果与试验值大为接近。从表2中可看出,使计算结果得到改进的主要因素是高压涡轮及动力涡轮的膨胀比因动力涡轮面积变化引起的改变量考虑得精确了,计入了导向器面积变化时$q^0(\lambda'_{c,a})$也随之变化(在亚临界下),从而部分地消减了面积变化影响能力的因素。试验装置中的动力涡轮导向器平均截面上的$\lambda_1$设计值为0.85。

表2 $F'_{c,a}$ 改变1%时的影响系数

| 性能参数 | 数据来源 | | |
|---|---|---|---|
| | 试验结果 | 文中计算结果 | 文献计算结果 |
| 压气机压比 | −0.48 | −0.31 | −0.50 |
| 燃气初温 | −1.05 | −0.65 | −1.24 |
| 高压涡轮膨胀比 | 0.59 | 0.54 | 1.12 |
| 动力涡轮膨胀比 | −1.01 | −0.84 | −1.59 |
| 功率 | −2.89 | −2.22 | −4.27 |
| 耗油率 | 0.83 | 1.08 | 2.03 |

转动动力涡轮导向器前后试验燃气轮机的外特性变化见图11。

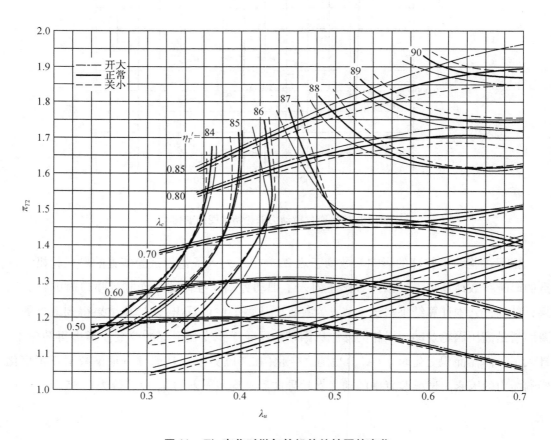

图11 $F'_{c,a}$ 变化时燃气轮机外特性图的变化

可以看到，开大动力涡轮导向器面积时，一方面输出功率下降，另一方面 $\eta_K$ 为常数时功率——转速线的峰值点向左移动。这是因为导叶开大时，若动力涡轮转速不变，则动叶入口气流角向负冲角增大方向变化，相应于最高涡轮效率的最大功率点处于更低的动力涡轮转速下。关小动力涡轮导向器面积时，动叶入口气流角向正冲角增大方向变化。因而外特性的功率峰值点向右移动。

虽限于结构因素，仅能进行动力涡轮导向器面积有限变动的试验验证，但文中所述方法可以讨论涡轮任一叶列面积变化对燃气轮机性能的影响。表3列出了试验燃气轮机的高压涡轮 I 级导向器、II 级导向器、动力涡轮导向器面积分别改变1%时对发动机性能影响的计算结果。诚如周知，各级导向器面积的影响能力及方向是不同的。例如，动力涡轮导向器面积对发动机功率、耗油率的影响较大，但在减小动力涡轮导向器面积使功率增大、耗油率降低的同时，燃气初温上升，工作点向压气机喘振边界方向移动。高压涡轮导向器面积对功率、耗油率的影响虽较前者为小，但在增大高压涡轮导向器面积使功率增大，耗油率降低的同时，工作点向远离压气机喘振边界方向移动，但亦受到燃气初温增加的限制。因此，在做性能调整时，可视对功率、耗油率、燃气初温、喘振裕度和部件工作寿命等诸因素的要求而加以综合利用。

**表3 试验燃气轮机导向器面积和动力涡轮效率的影响系数表（设计点）**

| | $\delta F_{c,a}$ | $\delta F_3$ | $\delta F'_{c,a}$ | $\delta \eta_{T_2}$ |
|---|---|---|---|---|
| $\delta \pi_{T_1}$ | -0.176 | -0.331 | 0.544 | 0.000 6 |
| $\delta \pi_K$ | -0.105 | -0.197 | -0.309 | -0.000 3 |
| $\delta \pi_\Sigma$ | 0.071 | 0.134 | -0.854 | -0.000 9 |
| $\delta \pi_{T_2}$ | 0.067 | 0.185 | -0.841 | 0.002 6 |
| $\delta \pi_\sigma$ | 0.005 | 0.009 | -0.013 | -0.003 5 |
| $\delta \pi_1$ | -0.486 | 0.189 | 0.038 | 0.000 0 |
| $\delta \pi_2$ | 0.030 | 0.249 | 0.050 | 0.000 1 |
| $\delta \pi_3$ | 0.128 | -0.725 | 0.210 | 0.000 2 |
| $\delta \pi_4$ | 0.151 | -0.043 | 0.247 | 0.000 3 |
| $\delta \pi'_1$ | 0.023 | 0.044 | -0.724 | 0.000 9 |
| $\delta \pi'_2$ | 0.043 | 0.081 | -0.117 | 0.001 7 |
| $\delta \pi_{1T}$ | -0.455 | 0.437 | 0.087 | 0.000 1 |
| $\delta \pi_{2T}$ | 0.279 | -0.768 | 0.457 | 0.000 5 |
| $\delta T_K$ | -0.030 | -0.056 | -0.087 | -0.000 1 |
| $\delta T_\Gamma$ | 0.080 | 0.151 | -0.646 | -0.000 7 |
| $\delta T_T$ | 0.119 | 0.223 | -0.764 | -0.000 8 |
| $\delta T_H$ | 0.104 | 0.196 | -0.582 | -0.119 8 |

表3(续)

| | $\delta F_{c,a}$ | $\delta F_3$ | $\delta F'_{c,a}$ | $\delta \eta_{T_2}$ |
|---|---|---|---|---|
| $\delta q(\lambda_{c,a})$ | -0.828 | 0.322 | 0.064 | 0.000 1 |
| $\delta q(\lambda_2)$ | 0.045 | 0.371 | 0.074 | 0.000 1 |
| $\delta q(\lambda_3)$ | 0.072 | -0.408 | 0.118 | 0.000 1 |
| $\delta q(\lambda_4)$ | 0.187 | -0.053 | 0.306 | 0.000 3 |
| $\delta q(\lambda'_{c,a})$ | 0.014 | 0.027 | -0.451 | 0.000 6 |
| $\delta q(\lambda'_2)$ | 0.035 | 0.066 | -0.095 | 0.001 4 |
| $\delta G_B$ | 0.026 | 0.049 | 0.077 | 0.000 1 |
| $\delta G_T$ | 0.219 | 0.411 | -0.011 | -0.001 2 |
| $\delta C_O$ | -0.048 | -0.089 | 1.083 | -1.005 2 |
| $\delta N_B$ | 0.267 | 0.500 | -2.220 | 1.004 0 |

从表3中亦可看出,导向器面积变动时对面积变动的本列、本级的膨胀比影响最剧,前级次之。面积变动对 $q(\lambda)$ 的影响量级也正说明了讨论导向器面积变动时应当计入 $q(\lambda)$ 变化的影响。

在导向器面积变动量较大时,应计入由此而引起的涡轮效率变化的影响。计算方法已同时提供了涡轮效率变化时的影响系数计算。作为例子,在表3中列入了计算得出的试验燃气轮机动力涡轮效率变化的影响系数。此时,面积变动的影响量应当和由导叶转角引起的效率变化的影响量叠加作为最终的影响值来加以讨论。

## 五、结束语

在燃气轮机的设计和调试实际中,改变涡轮导向器面积常常是一种有力的手段。

文中采用基于文献[1]基础上的小偏差法来讨论双轴燃气轮机(包含动力涡轮)中任一级涡轮导向器面积变化对燃气轮机性能的影响。在一台燃气轮机上进行的试验表明,所述方法改善了计算结果的精度,可在发动机的调整工作中加以采用。这种工程小偏差法计算简便,易于做一般分析,适于工程使用。该方法的一般适用范围是导向器面积改变量不超过 10% ~ 15%。

尽管限于结构因素,试验是局部的,但由于测量工作进行得较为系统,已基本展示了动力涡轮导向器面积变化对发动机性能影响的全貌。

## 参 考 文 献

[1] 契尔凯兹 AЯ. 航空涡轮喷气发动机原理和计算中小偏差法的应用[M]. 杨克立,译. 北京:国防工业出版社,1959.

# 附录 I  符 号 说 明

| | |
|---|---|
| $P_a, T_a$ | 大气压力、温度 |
| $P_B, T_B$ | 压气机前滞止气流的总压、总温 |
| $P_K, T_K$ | 压气机出口滞止气流的总压、总温 |
| $P_\Gamma, T_\Gamma$ | 涡轮前滞止气流的总压、总温 |
| $P_T, T_T$ | 高压涡轮后滞止气流的总压、总温 |
| $P_Д, T_Д$ | 动力涡轮前滞止气流的总压、总温 |
| $P_И, T_И$ | 动力涡轮后滞止气流的总压、总温 |
| $P_C, T_C$ | 排气管出口滞止气流的总压、总温 |
| $\pi_{T_1} = P_\Gamma / P_T$ | 高压涡轮膨胀比 |
| $\pi_K = P_K / P_B$ | 压气机增压比 |
| $\pi_\Sigma = P_Д / P_a$ | 动力涡轮入口压力与大气压力之比 |
| $G_B$ | 通过压气机的每秒空气流量 |
| $G_T$ | 燃烧室每小时的燃油消耗量 |
| $\pi_{T_2} = P_Д / P_И$ | 动力涡轮膨胀比 |
| $\pi_C = P_C / P_a$ | 排气管内外压力比 |
| $C_e$ | 发动机的耗油率 |
| $N_B$ | 发动机功率 |
| $q(\lambda_{c,a}), q(\lambda'_{c,a})$ | 高压涡轮、动力涡轮一级导向器的无因次密流 |
| $\pi_1, \pi_2, \pi_3, \pi_4$ | 高压涡轮逐圈膨胀比 |
| $\pi'_1, \pi'_2$ | 动力涡轮逐圈膨胀比 |
| $q(\lambda_2), q(\lambda_3), q(\lambda_4)$ | 高压涡轮第 2、3、4 圈无因次密流 |
| $q(\lambda'_2)$ | 动力涡轮第 2 圈无因次密流 |
| $\pi_{1T}, \pi_{2T}$ | 高压涡轮第 I、II 级膨胀比 |
| $\sigma_B = P_B / P_a$ | 进气道总压恢复系数 |
| $\sigma_\Gamma = P_\Gamma / P_K$ | 燃烧室总压恢复系数 |
| $\sigma_Д = P_Д / P_T$ | 中间扩压器总压恢复系数 |
| $\sigma_C = P_C / P_И$ | 排气管总压恢复系数 |
| $\eta_K$ | 压气机效率 |
| $\eta_{T1}, \eta_{T2}$ | 高压涡轮和动力涡轮的效率 |
| $\eta_m$ | 动力涡轮机械效率（或可包括减速器） |
| $q = 1 - \Delta G / G_B$ | 压气机末级空气相对抽气量 |
| $F_{c,a}, F'_{c,a}$ | 高压涡轮和动力涡轮 I 级导向器喉部面积 |
| $F_2, F_3, F_4$ | 高压涡轮第 2、3、4 圈喉部面积 |
| $F'_2$ | 动力涡轮第 2 圈喉部面积 |

$n_K, n_B$  压气机,减速器输出端转速

# 附录 Ⅱ  系 数 说 明

$K_1 = 0.286 \pi_K^{0.286} / (\pi_K^{0.286} - 1)$

$K_2 = (T_T - T_B) / T_K$

$K_3 = 0.25 / (\pi_{T_1}^{0.25} - 1)$

$K_4 = T_\Gamma - T_T / T_T$

$K_5 = T_\Gamma / T_\Gamma - T_K$

$K_6' = [0.125 / (\pi_C^{0.25} - 1)] - 0.75$

$K_{10} = \Delta \overline{G}_B \pi_{K_0} / \eta_{K_0} \Delta \pi_K$

$K_{11} = \Delta \eta_K \pi_{K_0} / \eta_{K_0} \Delta \pi_K$

$B_3 = 0.25 / \pi_{T_2}^{0.25} - 1$

$B_4 = T_T - T_H / T_H$

$A_2 = 1 / \left(1 - \dfrac{1}{2} B_3 B_4 + K_6'\right)$

$B_3' = 0.25 / \pi_{2T}^{0.25} - 1$

$B_4' = T_T - T_H' / T_H'$

$b_i = a_i \cdot K_{6i} \ (i = 1, 2, 3, 4)$

$i$ 为奇数时

$$a_i = \left[1 - \dfrac{\lambda_{wi}^2}{\lambda_{ei}^2} \cdot \dfrac{1}{1 - \dfrac{\tan(\beta_i - \alpha_i)}{\tan \beta_i}}\right]^{-1}, K_{6i} = \dfrac{K+1}{2K}\left(\dfrac{1}{\lambda_{ei}^2} - 1\right)$$

$i$ 为偶数时

$$a_i = \left[1 - \dfrac{\lambda_{ci}^2}{\lambda_{wi}^2} \cdot \dfrac{1}{1 - \dfrac{\tan(\alpha_i - \beta_i)}{\tan \alpha_i}}\right]^{-1}, K_{6i} = \dfrac{K+1}{2K}\left(\dfrac{1}{\lambda_{wi}^2} - 1\right)$$

$b_i' = a_i' \cdot K_{6i}' \quad i = 1, 2$

$i$ 为奇数时

$$a_i' = \left[1 - \dfrac{\lambda_{wi}'^2}{\lambda_{ci}^2} \cdot \dfrac{1}{1 - \dfrac{\tan(\beta_i' - \alpha_i')}{\tan \beta_i'}}\right]^{-1}, K_{6i}' = \dfrac{K+1}{2K}\left(\dfrac{1}{\lambda_{ci}'^2} - 1\right)$$

$i$ 为偶数时

$$a_i' = \left[1 - \dfrac{\lambda_{ci}'^2}{\lambda_{wi}^2} \cdot \dfrac{1}{1 - \dfrac{\tan(\alpha_i' - \beta_i')}{\tan \alpha_i'}}\right]^{-1}, K_{6i}' = \dfrac{K+1}{2K}\left(\dfrac{1}{\lambda_{wi}'^2} - 1\right)$$

$Z_i \approx Z_i' \approx 0.11$

# 船用燃气轮机箱装体板壁隔声设计 *

闻雪友

(哈尔滨船舶锅炉涡轮机研究所)

## 一、引言

随着舰船燃气轮机的不断发展和完善,其装舰数量在急剧增加。在发动机总体设计方面的一项重大改进,就是采用"箱装体"设计概念。所谓箱装体,就是将机组包容在一个完整而独立的密闭箱体和整体承力底座上。通常它包括燃气发生器,动力涡轮,进、排气管及发动机的全部附属设备。此外,还包括底架和机架及能承受冲击并具有良好隔热、隔音、抗核污染、抗生物和化学能力的箱体。

箱体有多种功能,其主要作用之一就是隔离和控制发动机产生的空气噪声。文中讨论如何更好地实现这一目标。

## 二、燃气轮机噪声源之声功率级

要控制燃气轮机的噪声,首先需了解噪声声源的噪声谱,然后再采取针对性的有效措施。燃气轮机在运行中有三个主要的噪声源:进气、排气和机匣辐射噪声。进气噪声主要与压气机级的压力、叶片数及转速成正比,特征是高频。机匣的辐射声取决于内部的噪声水平、机匣的几何形状及刚度。一般来讲,燃气轮机机匣的噪声级低于进、排气的噪声级(大多数燃气轮机的机匣噪声在低频和高频范围比排气和进气噪声低 10~15 dB),但也相当大,应作为一个主要的噪声源考虑。排气噪声的特征是宽带,其峰值一般向下移动到较低的频率范围。由于发动机周围各点到噪声源的距离不同、噪声源的能量大小不同以及来自不同噪声源的声波干涉、叠加的结果,能量级是不一样的。

发动机噪声谱数据可通过实测得到,但在设计阶段只能估算。

1. 首先求声功率级

$$PWL = 101 + 10\lg kw(\text{基准声功率 } 10^{-12} \text{ W},\text{下同})$$

式中,$kw$ 为发动机的功率,kW。

---

* 文章发表时间:1986 年 5 月。

由于系统的质量流量与燃气轮机功率成正比,故声功率级可用质量流量 $W(\text{lb/s})$ 表示:

$$PWL = 120 + 10\lg W$$

按功率或流量估算出声功率级后,可用表 1 中的修正系数求出倍频程声级。

表 1 倍频程修正系数(按燃气轮机排气噪声功率级)

| 倍频程/Hz | 37.5 | 75 | 150 | 300 | 600 | 1 200 | 2 400 | 4 800 |
| --- | --- | --- | --- | --- | --- | --- | --- | --- |
| | 75 | 150 | 300 | 600 | 1 200 | 2 400 | 4 800 | 9 600 |
| 修正系数/dB | -15 | -12.5 | -10 | -7.5 | -5 | -7.5 | -10 | -12 |

2. 进气噪声功率级

$$PWL = 109 + 10\lg kw$$

或

$$PWL = 127 + 10\lg W$$

可用表 2 中的修正系数求出倍频程声级。

表 2 倍频程修正系数(按燃气轮机排气噪声功率级)(dB)

| 倍频程/Hz | 37.5 | 75 | 150 | 300 | 600 | 1 200 | 2 400 | 4 800 |
| --- | --- | --- | --- | --- | --- | --- | --- | --- |
| | 75 | 150 | 300 | 600 | 1 200 | 2 400 | 4 800 | 9 600 |
| 用于 100 lb/s | -35 | -35 | -35 | -35 | -30 | -10 | -3 | -0 |
| 用于 200 lb/s | -30 | -30 | -30 | -30 | -25 | -10 | -5 | -8 |
| 用于 400 lb/s | -25 | -25 | -25 | -25 | -20 | -5 | -8 | -11 |

燃气轮机机匣辐射噪声数据列于表 3。

表 3 燃气轮机机匣辐射噪声(dB)

| 倍频程/Hz | 37.5 | 75 | 150 | 300 | 600 | 1 200 | 2 400 | 4 800 |
| --- | --- | --- | --- | --- | --- | --- | --- | --- |
| | 75 | 150 | 300 | 600 | 1 200 | 2 400 | 4 800 | 9 600 |
| 用于 100 lb/s | 115 | 116 | 117 | 112 | 108 | 108 | 107 | 119 |
| 用于 200 lb/s | 118 | 119 | 120 | 115 | 114 | 114 | 113 | 125 |
| 用于 400 lb/s | 121 | 122 | 123 | 128 | 120 | 120 | 119 | 131 |

由航空发动机改装的燃气轮机声功率级的估计可见表 4,其噪声特性与上相同。

按上述方法得到的噪声功率谱,与在一台由航空机改装的船用燃气轮机上所做的实测结果比较如图 1 所示。从图 1 可见,实测结果与表 4 的数值更接近些。

表4  由航空发动机改装的燃气轮机的噪声功率谱(dB)

| 倍频程/Hz | | 37.5 | 75 | 150 | 300 | 600 | 1 200 | 2 400 | 4 800 |
| --- | --- | --- | --- | --- | --- | --- | --- | --- | --- |
| | | 75 | 150 | 300 | 600 | 1 200 | 2 400 | 4 800 | 9 600 |
| 大致功率/kW | 大致流量/(lb/s) | 进气 | | | | | | | |
| 2 500 | 45 | 102 | 109 | 112 | 113 | 116 | 118 | 124 | 124 |
| 10 000 | 150 | 118 | 121 | 124 | 128 | 130 | 132 | 139 | 131 |
| 18 000 | 260 | 120 | 127 | 128 | 130 | 134 | 136 | 140 | 135 |
| 大致功率/kW | 大致流量/(lb/s) | 排气 | | | | | | | |
| 2 500 | 45 | 105 | 110 | 115 | 118 | 120 | 118 | 107 | 102 |
| 10 000 | 150 | 110 | 115 | 120 | 123 | 123 | 121 | 112 | 109 |
| 18 000 | 260 | 120 | 126 | 134 | 132 | 131 | 131 | 131 | 122 |
| 大致功率/kW | 大致流量/(lb/s) | 机匣 | | | | | | | |
| 2 500 | 45 | 101 | 101 | 101 | 106 | 110 | 110 | 111 | 114 |
| 10 000 | 150 | 114 | 112 | 114 | 116 | 120 | 120 | 126 | 121 |
| 18 000 | 260 | 116 | 127 | 120 | 120 | 124 | 124 | 127 | 125 |

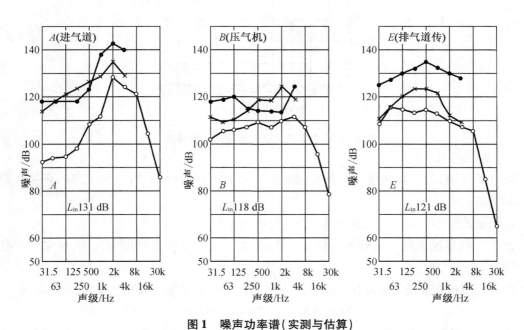

图1  噪声功率谱(实测与估算)

○——实测值　●——估算值　×——表4值

## 三、板壁设计的声学考虑

评价箱体声学性能的最方便指标是辐射能量衰减率,因为其不仅包括箱体壁面提供的能量传递损失,而且也计入了箱体固有的吸音特性。

辐射能量衰减率定义为

$$D = TL + 10\lg(S_a/S_w)$$

式中　$TL$——壁面传声损失;

　　　$S_a$——箱体内表面面积与其各自吸音系数的乘积之和;

　　　$S_w$——箱体外部辐射面积。

噪声的允许值以船员的听力不受危害为限。同时也要考虑对话清晰度的需要,以不影响有效的操作。因此,应对箱体的隔音性能提出要求。例如,美国通用电气公司对LM2500箱体隔音性能的要求是使辐射能量衰减率等于或大于图2所示之值。为确保最终定型的箱体能提供所需的辐射能量衰减率,最好先用试验板进行声学试验,以确定吸音系数和按倍频程计算的传声损失,然后再在样机上进行全尺寸的声学试验。

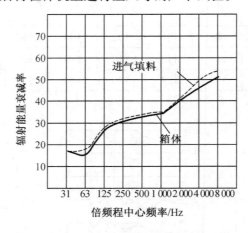

**图2　LM2500箱体辐射能量衰减率要求**

设计箱体时,为保证足够的辐射能量衰减率,声音传递损失和吸音系数之间必须达到适当的平衡。

1. 箱壁的结构应使声音传递损失最大(主要的考虑因素)

由于声音传递损失与箱壁的质量成正比,因此在结构质量允许的情况下,增加壁重有利于隔音。

单层均质板在质量控制范围内无规则入射条件下的质量定律经验公式为

$$TL = 18\lg m + 12\lg f - 25 \text{ dB}$$

式中　$m$——板壁的面密度,$kg/m^2$;

　　　$f$——频率,Hz。

平均隔声量(取 500 Hz 时的隔声量)的经验公式为($m < 100$ kg/m² 时)

$$\overline{TL} = 13.5\lg m + 13 \text{ dB}$$

从上两式可见,用单层板壁来实现高隔声量是很笨重的。

常采用双层壁。壁间为空气的隔声量计算式为

$$TL = 18\lg(m_1 + m_2) + 12\lg f - 25 + \Delta TL$$

修正项 $\Delta TL$ 随空气层的厚度增大而增加(图3)。图中曲线具有渐近性质,故应在尺寸、质量与经济性上适当平衡。其平均隔声量为($m_1 + m_2 < 100$ kg/m² 时)

$$\overline{TL} = 13.5\lg(m_1 + m_2) + 13 + \Delta TL$$

可见采用双层壁提高了隔声量。

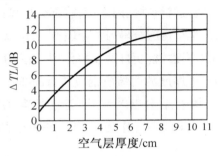

图3 修正项 $\Delta TL$ 与空气层厚度的关系

采用双层、不等厚结构,双层壁的隔声频率特性曲线上就不再出现吻合谷;而采用双层同种材料等厚结构,其吻合谷依然存在。

2. 箱壁内表面的吸声系数要尽可能大。

箱体的辐射能量衰减率不仅与各壁面的传声损失有关,而且与内表面所具有的吸声单位 $S_a$ 和箱体外部辐射面积 $S_w$ 有关。吸声单位 $S_a$ 可表示为

$$S_a = \alpha_1 S_1 + \alpha_2 S_2 + \cdots + \alpha_i S_i$$

式中 $\alpha_i$——材料的吸声系数;

$S_i$——吸声系数为 $\alpha_i$ 的吸声材料的面积,m²。

因此,吸声单位 $S_a$ 越大,则实际隔声量也就越大。

由于声源被罩在罩壳内部,声波在罩内经过来回多次反射,增加了混响声。如果罩壳内壁未进行很好的消声处理,则反射波声能很大,会导致声场内场压级的大幅度提高,这对隔声降噪不利。为了提高消声效果,内壁面可用微孔板,双层壁之间填充吸声材料。当穿孔率 $p \geq 0.2$ 时,内壁对吸声材料的吸声性能影响小,趋于稳定。

通常,填充吸声材料后效果可提高 3~10 dB。从分频来看,与不填充吸声材料相比,中、高频提高得多些,低频部分的效果提高得少些。通常情况下,填满柔软吸声材料要比不填满的效果好些,特别是填充物稍被压缩效果更好。为保证既有足够的吸声效果又比较经

济，其厚度一般为 50～75 mm。

需指出的是，填充的吸音材料必须用一种能承受箱体内部一定温度的薄膜包覆，如聚酰亚胺薄膜。这样可使吸音材料不受发动机运行环境中的油、气的渗透的影响，并使整个箱体内部都用低压淡水冲洗。然而，包覆薄膜有降低材料吸音特性的作用。此外，双层板之间的连接件构成声桥，会降低隔声量，设计时亦应予以注意。

3. 薄板型的箱体结构很容易因激振而形成再生性声源，因此可在箱体外壁的内表面涂敷减振消声阻尼剂。

通常消声减振阻尼涂料是由高分子树脂加入适量的填料以及辅助材料配制而成，是一种涂在各种金属板材表面上，具有减振、隔声、绝热和一定密封性能的特种涂料。在所涂敷的结构发生振动和噪音辐射时，能通过材料黏性内摩擦，将部分机械能转变为热能，从而达到减振和消声的目的。如果涂料与声学结构配合得当并合理施工，可以获得良好的减振和消声效果。

阻尼涂料的涂层需达一定厚度才能更好地发挥作用，因此在使用时应特别注意控制漆膜的厚度。一般涂层厚度约为基板厚度的两倍或钢板质量的20%左右。美国环境元件公司在为 LM2500 燃气轮机制造的箱体中，在外壁内表面上涂的是消声涂料公司的 GP-1 涂料。

4. 为达到箱体实际隔声量的要求，作为其构件的板壁的隔声量应高于箱体实际隔声量。

辐射能量衰减率的表达式亦可改写成

$$D = TL + 10\lg \alpha + 10\lg S - 10\lg S_w$$

因为吸声材料的吸声系数 $\alpha < 1$，因此式中 $10\lg \alpha$ 项是负值。$\alpha$ 很小时，该项是个较大的负值。对于实际箱体，$S < S_w$，因而 $(10\lg S - 10\lg S_w)$ 项亦是负值。再考虑到实际箱体结构中各构件间的连接及维修门等处可能有漏音，箱体上某些部位可能安设附件而局部影响隔声效果等。因此对板壁的隔声量要求应高于箱体的隔声量。

此外，在对板壁的隔声量进行估算时应考虑如下因素。

（1）如果是单层板结构，在按质量定律估算隔声特性时，应注意在临界频率处传声损失出现隔声低谷（吻合谷）。吻合谷随材料阻尼的减小而加深，在越过吻合谷之后，传声损失又以某个斜率上升，然后逐渐减缓，这又与质量控制时相一致。

临界频率可由下式计算：

$$f_0 = \frac{c^2}{2\pi}\sqrt{\frac{12m}{D}}$$

$$D = \frac{1}{12} \cdot \frac{Eh^3}{1-\mu^2}$$

式中　$c$——空气中的声速，m/s；

　　　$D$——板的劲度；

　　　$E$——板的弹性模量，N/m²；

$h$——板厚，m；
$\mu$——泊桑比。

（2）如果是双层中空结构，在共振频率下其隔声频率特性曲线上会出现一个低谷。共振频率可按下式计算：

$$f_0 = \frac{1}{\pi}\sqrt{\frac{\rho c^2}{(m_1+m_2)d}}$$

式中 $\rho$——空气的密度，kg/m³；
$d$——双层壁之间的距离，m。

若将空气之 $\rho$、$c$ 值代入，上式可写成

$$f_0 = \frac{120}{\sqrt{(m_1+m_2)d}}$$

5. 要从声学角度对板壁的刚度加以考虑。

壁的刚度与板材厚度、加强筋的形式及其间距等有关，也与对 31.5 Hz 和 63 Hz 倍频程的非常严格的要求有关。例如，在 LM2500 箱体设计时，要求无论总结构或主板中的顶板、侧板的自然频率都不得低于 80 Hz。

6. 在考虑板壁声学特性的同时，必须兼顾其他问题。

文中虽仅限于讨论箱体板壁的声学设计，但下列问题也必须统一考虑。

（1）箱体也是一个隔热件。箱体把燃气轮机大量的辐射热关闭在箱体内，因此对当量导热率有一定的要求。在考虑板壁结构和选择吸声材料、包覆时应予注意。LM2500 箱体设计成在 200 °F，在非构架区具有小于或等于 0.47 W/m²K 的当量导热率。

（2）箱体要承受规定的压差（包括内部负压和正压），满足规定的抗冲击性能。此时，箱体结构组件的最大工作应力不超过材料的屈服应力。箱体所需承受的环境条件还包括从 −40～+50 ℃ 的进气温度，可能高达 200 ℃ 的涡轮热辐射以及船舶在海上航行中甲板下机械受到的含盐湿空气，材料也必须能经受海军燃油的侵蚀，能允许在内部用低压淡水冲洗。这些也必须在声学设计时统一考虑。

（3）造价和总质量问题。

## 四、隔声试验

设计了一个 2 m×1 m 的隔声试件。主板为 6 mm 厚的钢板，面板为 1 mm 厚的微孔板或 1.5 mm 厚的钢板，主板与面板间距离为 80 mm，材料均为 A3 钢。充填的吸声材料为超细玻璃棉、硅酸铝板和酚醛纤维，均用聚酰亚胺薄膜包覆，再加上阻尼涂料，排列组合成九个试验方案。

试验在同济大学声学研究所的标准隔声试验室进行。由于试验中发现钢板主板与水泥边框连接处不严密等原因，试验结果虽未能真实地反映构件的实际隔声量，但所进行的

相对试验可以说明下列问题。

1. 同样的主结构用三种不同的吸声材料(硅酸铝板17.6 kg、酚醛纤维5.3 kg、超细玻璃棉5.6 kg)充填,其隔声性能的试验结果示于图4。由图4可见,超细玻璃棉的隔声效果最佳,其平均隔声量较前两者增加6.5~7 dB,而且用量小,价格也最便宜。

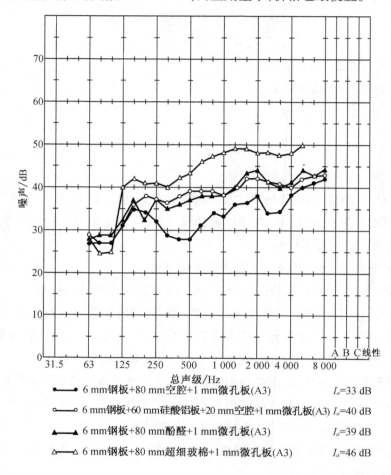

图 4  试验曲线

2. 从图4、图5可见,主板加微孔板结构充填吸声材料后,平均隔声量增加约11.7 dB。主板加光面板结构充填吸声材料后,平均隔声量增加约5.3 dB。显然,用微孔板能更好地发挥吸声材料的性能。

3. 从图6可见,在同样的主板加超细玻璃棉再加光面板结构的使用阻尼涂料的效果。

4. 从图7可见,如果不用吸声材料充填,则主板加光面板的结构比主板加微孔板的结构平均隔声量增加约3.2 dB。

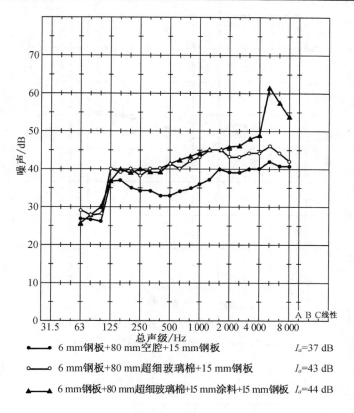

图 5 试验曲线

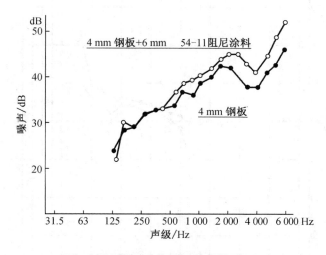

图 6 某船舶机房隔声试件隔声量比较

5. 图 8 反映了在结构完全相同的情况下,局部未充填超细玻璃棉(约 0.5 m² 的面积,占 25%)的影响,以模拟实际箱体中因装设附件而可能出现的情况,平均隔声量下降了约 2.3 dB。

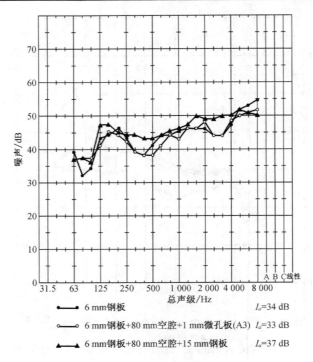

图7 试验曲线

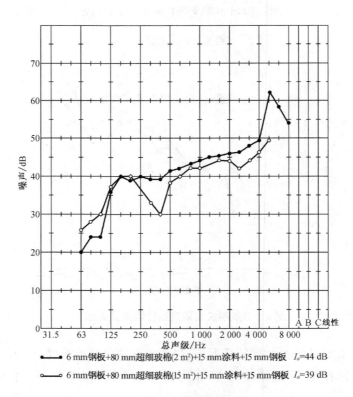

图8 试验曲线

## 五、结束语

文中讨论了船舶燃气轮机箱装体板壁声学设计的主要考虑方面,并通过试验强化了前述的讨论,可供设计参考。微孔板加聚酰亚胺薄膜,加超细玻璃棉,加消声阻尼漆,再加主板的结构模式,对所讨论的对象是适宜的。

试验工作委托同济大学声学研究所进行。参加试验工作的还有方长裕、牛文科和杨丽君同志。承倪乃琛副教授热情帮助,谨致谢意。

## 参 考 文 献

[1] 车世光.噪声控制与室内声学[M].北京:工人出版社,1981.
[2] 中国建筑科学研究院建筑物理研究所.建筑围护结构隔声[M].北京:中国建筑工业出版社,1980.
[3] 子安胜.建筑吸声材料[M].高履泰,译.北京:中国建筑工业出版社,1975.

# 带膜盘联轴器的轴系临界转速分析[*]

闻雪友

**摘　要**：文中从以典型轴系为中心展开的讨论中,得出了几个对于正确设计带膜盘联轴器的轴系(从临界转速方面)所应注意的结论。

## 一、引言

船舶燃气轮机装置上,通常是几个转子、转轴通过联轴等元件联成一个完整的轴系。严格地讲,只有轴系的临界转速才是实际的临界转速。鉴于发动机在临界转速工作的危害性,正确地设计轴系、精确地计算其临界转速,成为一个重要的问题。

近十几年来,膜盘式挠性联轴器被引入船用燃气轮机轴系中。图1为某燃气轮机输出轴示意图。

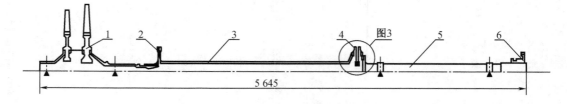

**图1　某大功率舰船燃气轮机的输出轴示意图**

1—动力涡轮；2—单膜盘联轴器；3—间隔轴；4—双膜盘联轴器；5—齿轴箱中的连接轴；6—离合器之部分

金属膜盘式挠性联轴器是一种通过极薄的双曲线型面的挠性盘来传递扭矩的装置。输入与输出轴向的相对位移由膜盘材料的挠性来吸收,膜盘可以单独,也可多个成系统应用。

---

[*] 文章发表时间：1979年10月。

膜盘联轴器在船舶燃气轮机上获得了应用,主要是由于其如下的突出性能:

(1)高速性、大容量,能以等速率传递扭矩。
(2)能吸收较大的相对位移;无须保养而又具有长寿命。
(3)无游隙、无噪音、无须润滑;长久不变的低不平衡量。
(4)作用在连接设备上的负荷小;能在恶劣环境下运行。

例如,美国的FT-4,LM2500等船用燃气轮机都在动力涡轮输出轴系中采用了膜盘式挠性联轴器。这些航空改装的大功率燃气轮机通常都带有长长的、越出排气管的动力涡轮输出轴,整个动力涡轮转子输出系统为一长达五六米以上的多跨转子。因此,临界转速问题较突出,而在这样的轴系中采用挠性膜盘联轴器,将使问题更为突出(图2)。必须在轴系中正确地采用、设计、布置膜盘联轴器。

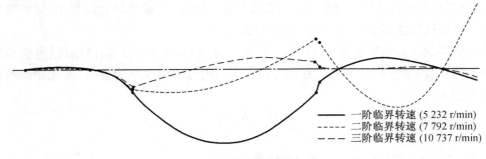

一阶临界转速 (5 232 r/min)
二阶临界转速 (7 792 r/min)
三阶临界转速 (10 737 r/min)

**图2  挠度曲线**

文中讨论由于在轴系中采用膜盘后,在临界转速计算及轴系设计方面所带来的若干新问题。为使叙述简明,讨论以一种典型的大功率燃气轮机的输出轴系(图1、图3)为例展开。该轴系的组成是:两端支承在滚动轴承上的双级动力涡轮转子;前端为单膜盘,后端为双膜盘的两个联轴器与间隔轴组成的挠性轴;齿轮箱中的连接轴,其后端带一离合器。

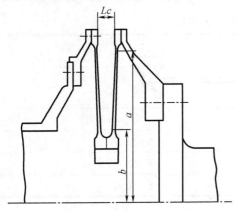

**图3  双膜盘联轴器详图**　　**图4  单个膜盘简化为一个具有一定弯曲弹性率的铰链**

## 二、膜盘联轴器的简化模型

文中讨论中临界转速计算采用 Prohl 提出的方法[1]。该方法是目前应用较普遍的方法之一，具有通用性与较简易性相结合的特点。这种方法对所给出的简化系统能给出一个精确解。对于真实转子而言，方法的精确性完全取决于表示真实转子及支承的简化系统和边界条件的严密程度。因此，对于带膜盘联轴器的轴系来说，问题在于如何对膜盘联轴器进行简化，以引入计算系统中。

1. 单膜盘的简化模型

膜盘是一种具有特定的弯曲弹性率的弹性元件。因此，严格地说，单个膜盘应简化为一个具有一定弯曲弹性率的铰链（图4）。但是注意到膜盘是作为一种挠性元件而用于轴系中的，因此其弯曲弹性率的设计值不应当是极高的。通常，在临界转速计算中，可把单膜盘简化为一个铰链连接而不必计入弯曲弹性率的影响。

表1所列的计算结果表明了这一点。计算是对轴系（图1）中的间隔轴段（并设该轴两端各带一单膜盘）进行的。由于在膜盘的工作范围内弯曲弹性率为常数，因此在两端的膜盘上的弯矩为

$$M_0 = K_{b_0}\theta_0$$
$$M_W = K_{b_W}\theta_W$$

式中　$K_{b_0}, K_{b_W}$——两端单膜盘的弯曲弹性率，$kg \cdot cm/(°)$；

　　　$\theta_0, \theta_W$——挠曲角，$(°)$。

从计算结果可见，随着 $K_b$ 值的增大，临界转速（$N_{cr1}$）也随之有所增高。但是比较设计值 $K_b = 20\,650\ kg \cdot cm/(°)$ 与 $K_b = 0$ 时的结果可知，略去 $K_b$ 的影响仅给临界转速计算带来极小的误差。

**表1　弯曲弹性度对临界转速的影响**

| $K_b/(kg \cdot cm/(°))$ | $Ncr_1/(r/min)$ |
|---|---|
| 0 | 6 652.7 |
| 20 650 | 6 653.2 |
| $10^7$ | 6 880.0 |

2. 双膜盘的简化模型

由于系统的惯性及弹性是影响系统振动的两个主要因素，因此在简化模型中，主要问题是如何折算当量质量及当量刚度，以反映实际系统的振动特性。

带双膜盘的膜盘联轴器等效轴表达式的导出条件如下:

(1) 膜盘联轴器折合前后该段的弯曲刚度应相等,即 $EJ = EJ_q$。

(2) 折合前双膜盘为一两端铰接轴,其刚性系数 $K_1 = EJ\pi^4/L_C^3$。折合为等效轴后,其与轴系的连接成为固持,其刚性系数 $K_1 = EJ_q \cdot \left(\frac{9}{4}\pi^2\right)^2 / L_q^3$。折合前后边界条件虽发生了变化,然其刚性系数应保持相等。

(3) 膜盘是一种具有特定弯曲弹性率的挠性元件,折合前后的弯曲弹性率应相等,因而 $K_b \cdot 90/\pi = EJ_q/L_q$。

最终可导出带双膜盘的膜盘联轴器的等效轴的表达式为

$$OD = (584 K_b L_q / E)^{1/4}$$
$$L_q = L_C \cdot \sqrt{3} \tag{1}$$

式中　$OD$——等效轴直径,cm;

　　　$L_C$——实际膜盘间隔(图3),cm;

　　　$L_q$——折合后的等效长度,cm;

　　　$E$——弹性模量,kg/cm$^2$。

例如,所讨论轴系中的双膜盘联轴器,膜盘的外径为 560 mm,$K_b$ = 20 650 kg·cm/(°),折合后为一直径 2.35 cm 的实心细轴。双膜盘轴器经折合成等效轴后即可方便地加入计算系统。

## 3. 带膜盘跨的简化模型

方案设计阶段需要通过单跨计算来选择轴的尺寸与跨度。实际上,轴系的临界转速与各单跨轴的临界转速是既有区别又有联系的,这一特点也常被用来在方案设计阶段根据单跨轴的临界转速粗略地估计轴系的临界转速,因为单跨轴的临界转速易于快速计算完成。

在带膜盘跨中,由于长长的挠性轴的质量全借两端的膜盘连接在相邻转子上,况且双膜盘联轴器的横向弹性率又很低,往往使带膜盘跨成为轴系各跨中临界转速的最低者。

在带膜盘跨的计算中,考虑到膜盘所具有的铰接特性,虽然其两端支承在滚动或滑动轴承上,却应将两端支承简化为固持。通过分析挠度曲线可明显看出这样简化较接近实际情况。诚然,可以预料,这样的简化结果将使所得计算结果高于实际值,应加以修正后作为轴系临界转速的估算值,即

$$N_{CR,P} = K \cdot N_{CR,T}$$

式中,系数 $K$ 主要取决于转子上与膜盘连接端的外伸段长度、膜盘数量以及相邻跨转子的结构形式,可取 $K = 0.85 \sim 0.95$。

图1轴系的计算实例示于表2。轴系一阶临界转速的精确值为 5 232 r/min,按上述原则简化计算的带膜盘跨的临界转速为 5 704 r/min,相应的系数 $K = 0.92$。

注意图2的挠度曲线,双膜盘联轴器确实表现出了双铰的特性,也可以看出其是如何影响临界转速的。

需指出,在对整个轴系作临界转速计算时,仍应将支承简化为铰支,这是显然的。

表 2　轴系临界转速计算结果(系统与各跨)

| 序号 | 轴系示意 | 临界转速/(r/min) | | | 备注 |
|---|---|---|---|---|---|
| | | 一阶 | 二阶 | 三阶 | |
| I | 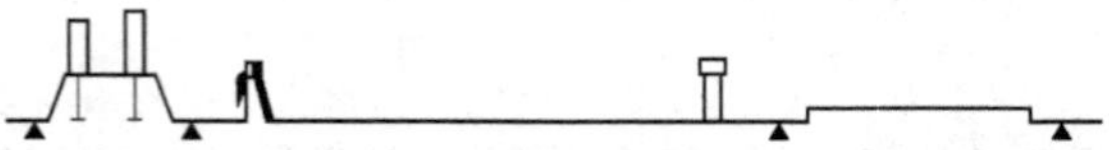 | 5 232 | 7 792 | 10 732 | |
| II | 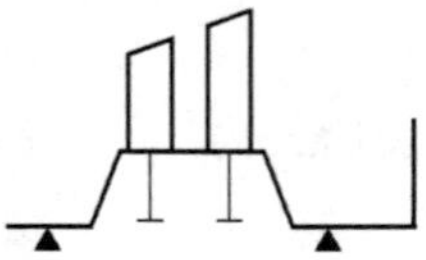 | 7 001 | — | — | 近似计入挠性轴质量的影响 |
| III |  | 7 348 | — | — | 近似计入挠性轴质量的影响 |
| IV |  | 5 704 | — | — | |
| | 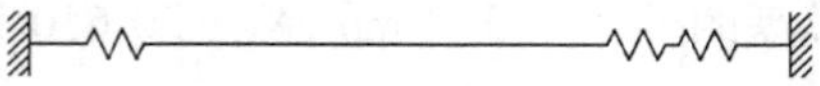 | 1 370 | — | — | |

## 三、膜盘特性对轴系临界转速的影响

膜盘联轴器中影响轴系临界转速的主要因素是:膜盘的弯曲弹性率 $K_b$ 以及双膜盘联轴器中膜盘间的间隔 $P_T L_C$。

应当指出,当讨论带膜盘联轴器的轴系临界转速时,挠性轴跨中的间隙轴是"悬浮"在两端的膜盘之间的,因此还涉及轴向振动频率问题。由于膜盘的弯曲弹性率 $K_b$ 与膜盘的轴向弹性率 $K_a$ 之间可以建立起如下的关系:

$$K_a = \frac{C_4 \cdot K_b}{2\pi a^2 (1+v) C_3}$$

式中　$K_a$——膜盘的轴向弹性率,kg/cm;

　　　$C_3, C_4$——系数,根据 $b/a$ 计算得出;

　　　$a, b$——膜盘的计算外半径与内半径(图3);

　　　$v$——泊桑比。

因此,当讨论 $K_b$ 变化对临界转速的影响时也必须涉及因 $K_a$ 变化造成的对挠性轴轴向振动频率的影响。虽然,通常认为轴向振动问题不太严重,但是针对轴向振动,在轴系中采用专门的结构措施的实例也并不乏见。

1. 膜盘弯曲弹性率 $K_b$ 对临界转速的影响

单膜盘的弯曲弹性率主要取决于膜盘的厚度、$b/a$ 以及材料本身的特性。单膜盘之 $K_b$ 对临界转速的影响前已涉及,这里仅讨论双膜盘联轴器的情况。

利用等效轴的表达式,把 $K_b$ 的变化转化为等效轴当量直径的变化。显然,膜盘弯曲弹性率的减小将使轴系的临界转速下降。对图 1 轴系的计算结果如图 5 所示。

应当注意,图 5 中一阶临界转速的这种变化趋势:当 $K_b$ 值过小时,随着 $K_b$ 值的减小,临界转速值的下降加速。由于大功率船用燃气轮机通常都具有长长的挠性轴(尤其是由航空发动机改装成的船用发动机),在结构设计上总是设法要提高轴系的临界转速,因此单膜盘的 $K_b$ 值不宜设计在使临界转速下降较剧烈的区域内。

$K_b$ 值变化引起图 1 轴系中挠性轴轴向振动频率变化的情况见图 6。$K_b$ 值的变化同时也引起作用在连接件上负荷(力和力矩)的变化。

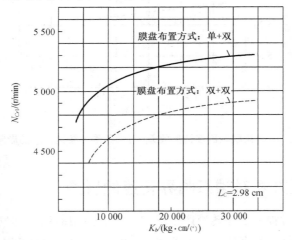

图 5　膜盘的弯曲弹性率 $K_b$ 对轴系临界转速 $N_{cr1}$ 的影响

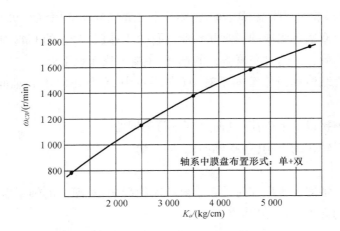

图 6　单膜盘轴向弹性率 $K_a$ 对轴系轴向振频 $\omega_{CR}$ 的影响

## 2. 膜盘间隔 $L_C$ 对临界转速的影响

可直观地看出,膜盘间隔 $L_C$ 增大将使轴系临界转速下降。

从双膜盘联轴器的折合横向弹性率表达式上可做出解释,即

$$K_d = \frac{360 K_b}{\pi L_C^2}$$

式中,$K_d$ 为联轴器折合横向弹性率,kg/cm。随着 $L_C$ 增加,$K_d$ 将急速下降。

上式已表明 $L_C$ 对轴系临界转速的影响比 $K_b$ 更剧,因此,为提高临界转速,在膜盘联轴器的结构设计中 $L_C$ 值应尽可能地保持最短。

对图1轴系的计算结果如图7所示。

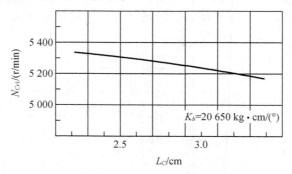

**图7　膜盘联轴器结构参数 $L_C$ 对轴系临界转速 $N_{Cr1}$ 的影响**

## 四、膜盘布置方式对轴系临界转速的影响

膜盘布置方式会对轴系临界转速产生影响。实际使用的船舶燃气轮机中,膜盘联轴器在轴系中的布置形式是多种多样的。举例来说,DDH-280级驱逐舰上的FT-4燃气轮机,动力涡轮端为单膜盘,减速器端为套齿连接;"欧洲班轮"号上的FT-4燃气轮机,动力涡轮端为单膜盘,减速器端为双膜盘;DD963级驱逐舰上的LM2500燃气轮机,在动力涡轮及减速器端各有一个由双膜盘组成的膜盘联轴器;而在"卡拉汉"号上的LM2500燃气轮机,仅在动力涡轮端采用双膜盘。这种布置方式的选择,主要取决于以下原因。

1. 对发动机组装体和齿轮箱间不对中(角不对中、平行不对中和轴向不对中)能力的要求。
2. 对作用在连接设备上负荷(轴向力、弯矩)的限制。
3. 对轴系临界转速的影响(包括轴向振动问题)。
4. 环境条件。

为讨论膜盘联轴器布置形式对轴系临界转速的影响,对图1的轴系做了各种布置形式下轴系临界转速的计算,其结果如表3。

表3 膜盘联轴器布置形式对临界转速的影响

| 序号 | 轴系示意 | 临界转速/(r/min) | | | 备注 |
| --- | --- | --- | --- | --- | --- |
| | | 一阶 | 二阶 | 三阶 | |
| Ⅰ | | 5 232 | 7 792 | 10 732 | 类似"欧洲班轮"上FT-4布置 |
| Ⅱ | | 4 824 | 7 455 | 9 634 | 类似DD963上LM2500布置 |
| Ⅲ | | 4 879 | 7 597 | 9 971 | 类似"卡拉汉"号上LM2500，涡轮端为双膜盘 |
| Ⅳ | | 5 362 | 7 823 | 11 101 | 类似DDH-280上FT-4及FT-12，涡轮端单膜盘，减速器端为套齿 |

所得结果与预料趋势一致：两端各带一单膜盘的轴系临界转速最高；两端各带一双膜盘的轴系临界转速最低；单、双膜盘组合的，其临界转速值居间。这些结论容易从双膜盘联轴器的高挠性上得到解释。需要指出的是，Ⅰ，Ⅱ两个方案的比较，同样是单、双组合的方案，然而二者的一阶临界转速却有较大的差别(7.2%)。通过比较Ⅱ，Ⅰ及Ⅲ，Ⅳ方案，可发现，在齿轮箱端采用相同的膜盘联轴器的情况下，将动力涡轮端由双膜盘改为单膜盘，一阶临界转速分别提高8.5%及9.9%。再比较Ⅰ，Ⅳ及Ⅱ，Ⅲ方案，可发现，在涡轮端采用相同的联轴器的情况下，将齿轮箱端的联轴器由双膜盘改为单膜盘，一阶临界转速仅分别提高了2.5%及1.1%，其影响远不如前者显著。这表明，在挠度较大的位置上布置双膜盘联轴器会使轴系临界转速较之在此位置上采用单膜盘联轴器时有较大的下降。反之，在挠度较小的位置采用双或单膜盘联轴器则对轴系临界转速影响较小。这一特点在需要兼顾临界转速和所要求的不对中能力时可加以利用。这实际下也说明了膜盘联轴器的布置应靠近支承，减少转子外伸长度，以提高轴系的临界转速。计算实例表明，这是一种有效的措施。

膜盘的布置方式显然也将影响轴系中挠性轴的轴向振动频率以及作用在连接件上的轴向力。计算结果示于图8和图9。

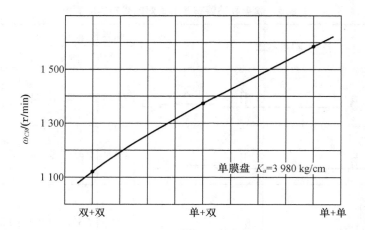

**图 8　轴系中膜盘布置形式对轴向振频的影响**

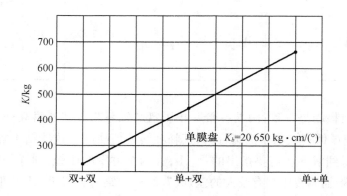

**图 9　轴系中膜盘布置形式对输出轴加在连接件上轴向力的影响**

## 五、支座弹性对轴系临界转速的影响

上述讨论中的计算都假定支座是绝对刚性的。实际上轴承座、基础和滑动轴承中的油膜均是弹性体。根据振动理论,轴承座的动刚度系数 $K$ 可由下式决定:

$$K = C_S - m_S \omega^2$$

式中　$C_S$——轴承座的静刚度系数,kg/cm;

　　　$m_S$——轴承座及基础参与振动的质量,kg·s²/cm;

　　　$\omega$——轴的角速度,1/s。

考虑油膜刚度支座的总刚度系数 $C$(kg/cm)可表示为

$$C = P \cdot \frac{C_S - m_S \omega^2}{C_S + P - m_S \omega^2)}$$

式中,$P$ 为油膜的刚度系数,kg/cm。

显然，动刚度系数 $K$ 和总刚度系数 $C$ 也都是随转速而变的。这些系数只能由实验确定。因此，这里仅是人为地对所讨论的轴系给定刚度系数，从而观察其对临界转速的影响趋势。根据轴承的类型分成两组：涡轮端的滚动轴承支座以 $\delta_1$ 给定；齿轮箱端的滑动轴承支座以 $\delta_2$ 给定。

图 10 表示 $\delta_1$，$\delta_2$ 相等并变化时对临界转速的影响。图 11 示出了 $\delta_1$ 为常数时，临界转速随 $\delta_2$ 而变的情况。图 12 则是 $\delta_2$ 为常数时，临界转速随 $\delta_1$ 变化的情况。可以看出，对于所讨论的轴系，当 $\delta_1$，$\delta_2$ 小于 $10^6$ kg/cm 时，临界转速开始显著下降。

膜盘联轴器能适应较大不对中的特点，常被用于发动机组装体通过减震器安装在基础上的结构场合。此时情况远较上述的把轴承座简化为一个自由度的弹簧——质量系统复杂。因此，在按刚性支承计算、确定其所需设计裕度时，应考虑到支承及其座结构的影响。

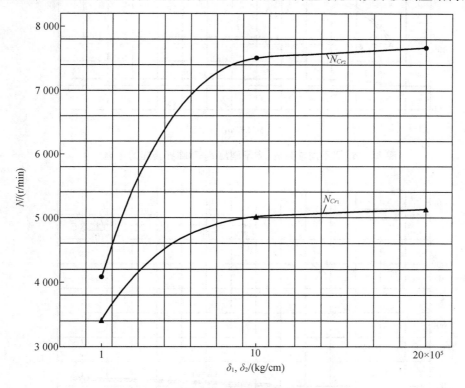

图 10　支承刚度对临界转速的影响（$\delta_1 = \delta_2$）

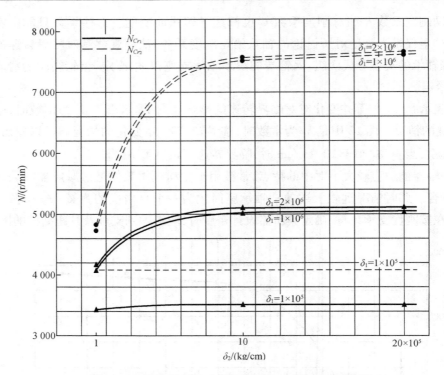

**图 11** $\delta_1$ 取不同常数时,支承刚度 $\delta_2$ 对临界转速的影响

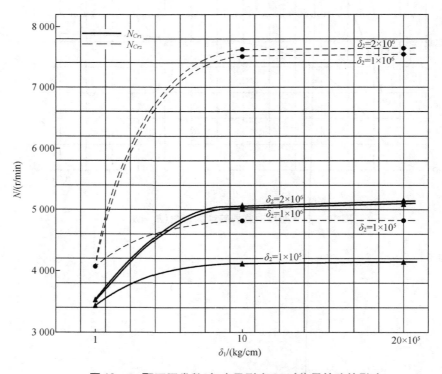

**图 12** $\delta_2$ 取不同常数时,支承刚度 $\delta_1$ 对临界转速的影响

## 六、结论

通过对若干带膜盘联轴器的轴系临界转速的数十个方案计算、比较、分析,得出下列结论供设计此类轴系时参考。

(1)对轴系进行临界转速计算时,单膜盘联轴器可简化为铰链,双膜盘联轴器可折合成等效轴。

(2)在对带膜盘跨进行方案估算时,应将支承边界条件处理为固持。

(3)从临界转速角度看,膜盘的弯曲弹性率应选择在对临界转速影响较平坦的区域,不宜过低。膜盘间的间隔距离 $L_c$ 应设计成尽可能地小。

(4)轴系中膜盘的布置应靠近支承,减少转子外伸段长度。在轴系挠度较大的位置上,联轴器由双膜盘改为单膜盘会在提高临界转速上有较大收益。

通常的轴系设计要求,在考虑到上述意见时,会有助于设计出一个从临界转速观点来看也较合理的带膜盘联轴器的轴系。

## 参 考 文 献

[1] PROHL M A. A general method for calculating the critical speeds of flexible rotors[J]. Journal of Applied Mechanics, Trans. ASME, 1945, 12(3): A142 – A148.

# 三、ASME 上发表的文章

# Feasibility Study of an Intercooled-cycle Marine Gas Turbine*

WEN Xueyou, XIAO Dongming

(Harbin Marine Boiler & Turbine Research Institute)

**ABSTRACT**

From the perspective of an overall entity analyzed was the performance obtainable from the adoption of an intercooled-cycle gas turbine under different typical cycle parameters of a gas turbine. On this basis, a study was conducted of the conversion of a high-power simple-cycle marine gas turbine (MGT-33) into a type of intercooled-cycle marine gas turbine. The precondition of the conversion is to keep the flow path and the majority of the structure of the original engine gas-generator unchanged in order to inherit the reliability of the prototype machine.

The results of the study indicate that after the adoption of an intercooled cycle under the precondition of performing minimum structure modifications and maintaining the compactness of the engine as a whole, there is still a significant enhancement of the gas turbine overall performance with its power output and efficiency being increased by about 34% and 4.6% respectively, demonstrating the merits of the engineering conversion under discussion.

**NOMENCLATURE**

$Ga$    flow rate
$LHV$    fuel lower heating value
$Ne$    power output
$P$    pressure
$PR_LPC$    pressure ratio of low pressure compressor
$PR_{HPC}$    pressure ratio of high pressure compressor
$PR$    total pressure ration
$T$    temperature
$TIT$    turbine inlet temperature
$SFC$    specific fuel consumption
$SP$    specific power output
$\eta$    polytropic efficiency
$\xi$    pressure loss coefficient, $\xi = \Delta P/P$
$\varepsilon$    intercooler efficiency, $\varepsilon = (T_2 - T_{22})/(T_2 - T_{cool})$

---

*本文为2007年美国机械工程师学会(ASME)燃气轮机年会论文。

$\eta_e$　cycle efficiency
$\Delta P$　pressure loss

**SUBCRIPTS**

in　inlet
LPC　low pressure compressor
IC　intercooler
HPC　high pressure compressor
CC　combustion chamber
HPT　high pressure turbine
LPT　low pressure turbine
PT　power turbine
ID　intermediate diffuser between low pressure turbine and power turbine
ex　exhaust
cool　coolant
0,1,2,3...　calculation points(see Fig. 2)

## Preface

Beginning from the late nineteen eighties and with the continually rising demand of main propulsion plants for surface naval vessels by navies around the world, high-power marine gas turbines(above 25 MW) have become the main current of development for the navies of various countries. For details, see Table 1.

**Table 1　General situation of the development of high-power marine gas turbines**
**(up to the year of 2005)**

| Model | LM2500 | FT8 | UGT-K25000 | WR-21 | LM2500 + | MT-30 |
|---|---|---|---|---|---|---|
| 1st year available | 1969 | 1990 | 1993 | 1997 | 1998 | 2001 |
| ISO MAX/hp | 33 600 | 36 860 | 42 400 | 33 850 | 40 500 | 48 275 |
| Cycle efficiency | 0.372 | 0.389 | 0.381 | 0.421 | 0.391 | 0.398 |
| SFC/kg/(kW·h)) | 0.227 | 0.217 | 0.221 | 0.200 | 0.215 | 0.212 |
| Pressure ration | 19.3 | 18.8 | 21.0 | 16.2 | 22.2 | 24.0 |
| Flow/(kg/s) | 70.4 | 83.3 | 87.6 | 73.1 | 85.8 | 116.7 |
| Manufacturer | GE | P&W | Zorya-mashporekt | R.R | GE | R.R |

\* Date taken from 2005 GTW Handbook.

Fig. 1 shows the development tendency of high-power marine gas turbine performance for the last 15 years. With a gradual increase in single-machine capaticy the maximum power is currently about 36 MW(50 000 hp at ISO conditions). There is also a gradual enhancement in efficiency with the efficiency of a simple cycle attaining 40% and that of a complex cycle 42%.

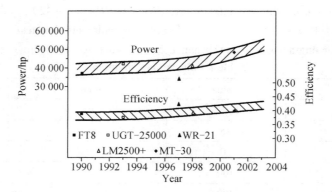

**Fig. 1  Trend of the change in power output and efficiency of high-power marine gas turbines**

It is generally recognized that marine gas turbines of 30 – 36 MW (40 000 – 50 000 hp) can meet the main propulsion plant requirements of high-power gas turbines for large and medium-sized surface vessels by the navies of various countries in the forthcoming 10 – 15 years.

There are two approaches for enhancing the performance of marine gas turbines: 1. On the basis of a traditional simple cycle by increasing the pressure ratio and turbine inlet temperature and by improving component efficiency it is possible to attain the desired new aim. In this connection, LM2500 + can serve as a representative engine; 2. by employing a complex cycle and through the introduction of cycle improvements a higher performance can be realized. In this case, WR-21 engine with the use of a complex cycle featuring intercooling and recuperation can be cited as a representative machine.

On the basis of an existing engine with its core components remaining maximally unchanged the authors have explored the possibility of developing a high-power marine gas turbine through the use of only a relatively simple intercooled cycle (IC).

## Intercooled cycle analysis

First, the relationship between the performance of an intercooled-cycle gas turbine and main cycle parameters is to be quantitatively examined. The engine is composed of a dual-shaft gas generator (including an intercooler) and a power turbine (See Fig. 2). The assumed component performance parameters are listed in Table 2.

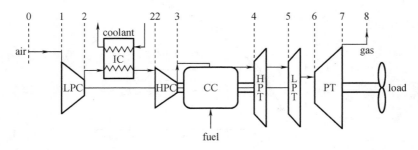

**Fig. 2  IC cycle schematic drawing**

Under the condition of different total pressure ratios ($PR = 12 - 42$), different turbine inlet temperatures ($TIT = 1\,100 - 1\,400$ ℃) and different intercooler efficiencies ($\varepsilon = 0 - 0.85$)[2] the calculation results for engine overall performance are shown respectively in Fig. 3 – Fig. 6.

**Table 2 Calculation Condition**

| parameter | Value | Parameter | Value |
|---|---|---|---|
| $P_0$/bar | 1.013 | $\eta_{CC}$ | 0.99 |
| $T_0$/℃ | 15 | $\eta_{HPT}$ * | 0.87 |
| $T_{cool}$/℃ | 20 | $\eta_{LPT}$ | 0.89 |
| $\Delta P_{in}$/mmH$_2$O | 100 | $\eta_{PT}$ | 0.92 |
| $\Delta P_{ex}$/mmH$_2$O | 400 | $\xi_{IC}$ | 0.05 |
| $PR_{LPC}/PR_{HPC}$ | 1 | $\xi_{CC}$ | 0.05 |
| $\eta_{LPC}$ | 0.88 | $\xi_{ID}$ | 0.01 |
| $\eta_{HPC}$ | 0.90 | $LHV$/kJ/kg | 42 700 |

\* With 1 200 ℃ serving as a benchmark, an increase (or decrease) of turbine inlet temperature by every 100 ℃ will lead to a decrease (or increase) in turbine efficiency by 1.25%.

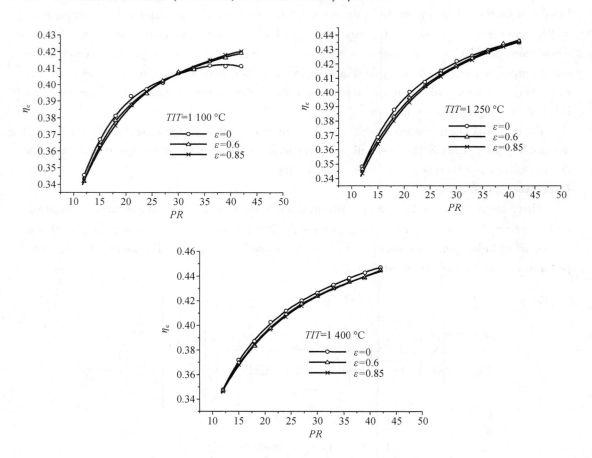

Fig. 3 Impact of intercooling on engine efficiency

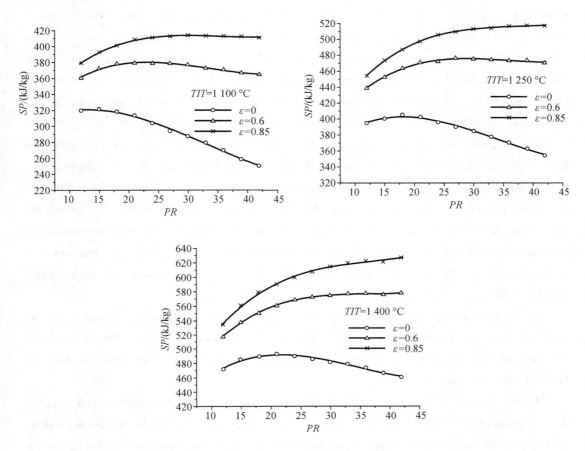

Fig. 4　Impact of intercooling on engine specific work

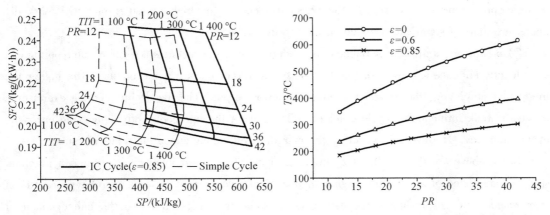

Fig. 5　Overall performance curves of a simple cycle and an intercooled one

Fig. 6　Impact of intercooling on compressor outlet temperature

1) From Fig. 3 one can perceive the impact of adopting intercooling on engine efficiency. Under the parameters of modern marine gas turbines(turbine inlet temperature at about 1 300 ℃,

pressure ratio 20 − 25) the use of intercooling has very little influence on engine cycle efficiency. Moreover, with an increase in intercooler efficiency there will be a slight decrease in cycle efficiency. This is because with the use of intercooling though the power consumption of the compressor is reduced and effective power output increased, there is a reduction in compressor outlet temperature. To make the gas achieve the predetermined turbine inlet temperature, one has to consume more fuel. Only under the condition of a high pressure ratio will the impact of a comperessor power consumption decrease surpass the influence of fuel increase caused by the introduction of intercooling. In this case its efficiency will be higher than that of a simple cycle.

2) Fig. 4 shows the impact of intercooling on engine specific work. It is obvious that the use of intercooling will lead to a significant increase in gas turbine specific work. The greater the intercooler efficiency, the greater the specific work. In the meantime, the pressure ratio value corresponding to a maximum specific work also increases dramatically. Under the parameters of modern marine gas turbines the specific work after the use of intercooling can be increased by 22% − 30%.

3) Fig. 5 shows the superimposition of the overall performance curves of a simple cycle and an intercooled one. It can be seen clearly that after the use of intercooling both the pressure ratio value corresponding to a maximum specific work or that corresponding to an optimum efficiency have all increased very significantly.

4) Fig. 6 shows the impact of intercooling on compressor outlet temperature. This signifies that the temperature of the cooling air entering the high temperature turbine blades has decreased. When the metal surface temperature of the high temperature turbine blades is kept unchanged, it is allowed to moderately enhance the turbine inlet temperature. In addition, when the compressor specific speed remains unchanged, the physical speed of the high pressure rotor will markedly decrease, thereby significantly reducing its working stress.

5) to sum up, under the parameters of modern marine gas turbines the use of intercooling can significantly enhance specific work and increase the power output of the engine. The higher the intercooler efficiency, the greater the effective power output. The use of intercooling exercises a relatively weak influence on cycle efficiency. The greater the intercooler efficiency, the lower the cycle efficiency. However, in general, the margin of efficiency change is relatively small.

Given above are the results of a simple analysis of an intercooled cycle. When a practical study is conducted of an existing engine involving its development to form a prototype engine of intercooled cycle gas turbine, one can discover some factors favorable to the enhancement of efficiency. Such factors include a moderate increase of turbine inlet temperature, an increase in working point efficiency of some components and the changes in specific speed of some components after their renewed matching, etc. The foregoing makes it possible to realize a simultaneous enhancement of engine power and engine cycle efficiency.

## Some specific case of intercooled cycle shemes

It is determined to perform a study by selecting a certain model of engine (MGT-33) to serve as a prototype engine and develop it to form an IC cycle. The aim of the study is to determine the technical feasibility of the selected scheme.

For the main performance of simple cycle gas turbine MGT-33, see Table 3.

**Table 3  Gas turbine MGT-33 performance at ISO conditions**

| Ser. No. | Parameter | Symbol | Unit | Value |
|---|---|---|---|---|
| 1 | Power | $Ne$ | kW | 28 500 |
| 2 | Cycle efficiency | $\eta_e$ | % | 37 |
| 3 | Air flow rate | $Ga$ | kg/s | 90 |
| 4 | Turbine inlet temperature | $TIT$ | K | 1 543 |
| 5 | Total pressure ratio | $PR$ | — | 22.2 |

The main principles governing the development of gas turbine MGT-33 to CGT-IC cycle gas turbine can be given as follows:

1) At ISO conditions CGT-IC has the same turbine inlet temperature as the prototype engine or an increase of the inlet temperature not exceeding 30 ℃;

2) At ISO conditions the specific speed of the CGT-IC low-pressure compressor is the same as that of the prototype engine;

3) Adopt to a maximum degree the flow path of the gas generator of the prototype engine;

4) The cooling structure of the high and low pressure turbine blades and also the high pressure turbine rotating blades will remain unchanged;

5) Inherit to a maximum degree the reliability of the prototype engine components.

The implementation of the above-mentioned principles has provided the basic preconditions for enhancing the power output and cycle efficiency of the engine after its modification. Moreover, under the precondition of making minimum changes to the structure of the gas generator of the prototype engine and by taking advantage of the maximum versatility of the flow path, the inherent high reliability can be readily taken over in its entirety.

For the cycle performance after optimization calculations, see Table 4. For the schematic drawings of MGT-33 and CGT-33, see Fig. 7.

**Table 4  Performance of gas turbine CGT-IC at ISO conditions**

| Ser. No. | Parameter | Symbol | Unit | Scheme I | Scheme II |
|---|---|---|---|---|---|
| 1 | Power | $Ne$ | kW | 36 400 | 38 200 |
| 2 | Cycle efficiency | $\eta_e$ | % | 38.4 | 38.7 |
| 3 | Air flow rate | $Ga$ | kg/s | 89.6 | 89.6 |
| 4 | Turbine inlet temperature | $TIT$ | K | 1 543 | 1 572 |
| 5 | Total pressure ratio | $PR$ | — | 22.3 | 21.2 |

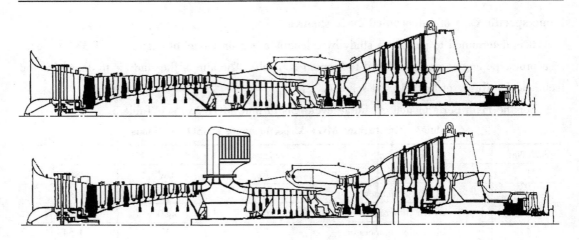

**Fig. 7 Schematic drawings of MGT-33 and CGT-33**

The technical highlights of the scheme (in our case, Scheme II) can be given as follows.

1) Compressor

From the low-pressure compressor the last two stages are to be removed with the new last stage guide vanes and straightening blades being designed anew. This enables the working points of the low-pressure compressor after stage removal and under the condition of keeping the reduced flow rate and speed unchanged to have the outlet pressure lowered and to balance the influence of the decrease of a reduced flow rate at the high-pressure compressor inlet cuased by the use of intercooling. The pressure ratio of the low-pressure compressor after adjustments will be reduced by 9.1% and the efficiency enhanced by 1.82%.

With the high pressure compressor being kept unchanged, the specific speed after a matching process will decrease by 0.2%, physical speed decrease by 7.3%, pressure ratio increase by 10.5% and efficiency decrease by 1.4%.

The compressor total pressure ratio will decrease by 4.5%. Apart from the above-mentioned causes, the pressure loss of the intercooler also plays a role.

2) Intercooler

The intercooler is a key component for raising power output by a large margin. The insertion of an intercooler between two compressors enables air to be pre-cooled before entering the high pressure compressor, thereby decreasing the power consumption of the high-pressure compressor and effectively enhancing the power output of the engine. To realize an intercooler with high compactness and low flow resistance, a modularized structure has been adopted. With an intercooler effciency of 0.85, the total pressure loss coefficient of the intercooler is 0.05.

3) High and low pressure turbine

Owing to the specific features of the prototype engine, such as a relatively high turbine inlet temperature, a complicated blade cooling structure of the high and low pressure turbine, the cooling structure of rotating blades of the high and low pressure turbine have been kept unchanged in order to lower the technical risk to a minimum. The structure of rotating blades of the high pressure turbine is kept unchanged. After a renewed matching performed to the high pressure turbine its expansion ratio is reduced by 25.6% and efficiency enhanced by 2.6%. This has taken place because of the significant decrease of the power consumption of the high pressure compressor.

The removal of the last two stages of the low pressure compressor has led to a reduction of the expansion ratio of the low pressure turbine by 9.7%, with its efficiency being kept unchanged after redesign. Moreover, owing to the decrease of the high pressure turbine expansion ratio and the increase of turbine inlet temperature, the temperature of the gas entering the low pressure turbine has increased by 7% (85 ℃).

Due to a significant decrease in the temperature of the cooling air entering the high and low pressure turbine blades the metallic surface temperature of the blades will not exceed the corresponding temperature of the prototype engine. Moreover, the lowering of the physical speed of the high pressure rotor has led to a decrease in centrifugal force by 31.9%, but the increase of thermal stresses in blades should be taken into account.

4) Power turbine

The use of intercooling and the removal of the last two stages from the low pressure compressor have resulted in the decrease of power consumption by the high and low pressure compressor, an increase in power turbine inlet temperature by 9.2%, an enhancement of expansion ratio by 42.3%, a significant increase in power-turbine power output, and a need to design the power turbine anew. The efficiency of the power turbine designed anew has increased by 1%, leading finally to a power output increase amounting to 34%. As the efficiency of the power turbine is by far higher than that of the high and low pressure turbine, a considerable portion of energy has been transferred to a high-effective zone for expansion and doing work. This has played a helpful role in improving the efficiency of the engine as a whole. Hence, so far as an IC cycle is concerned, it is especially important to design a high-efficiency power turbine.

5) Critical speed of the low pressure rotor

Due to the necessity of inserting an intercooler between the high pressure and low pressure compressor of the prototype engine, it is inevitable to cause an increase of the axial dimensions of the low pressure rotor, leading to a change in the critical speed of the low pressure rotor. One of the reasons for the removal of the last two stages from the low pressure compressor consists in decreasing the influence of the increase in axial dimensions of the low pressure rotor caused by the introduction of an intercooler, and also in controlling the engine total length.

6) Combustor and fuel oil system

The use of intercooling will cause a significant decrease of the temperature of air entering a combustor, but the turbine inlet temperature will not change. Hence, fuel consumption will increase (relative to simple cycle, the increase being assessed at 28.3%). As for fuel oil pumps, fuel oil spray nozzles, combustor and fuel oil regulation system there arises a need for performing pertinent adjustments or a renewed design.

## Performance comparison

For CGT-IC gas turbine power output and oil consumption performance curves, see Fig. 8. For comparison purposes also given are the prototype engine performance and calculated performance for developing MGT-33 to CGT-ICR(used only for conducting comparisons).

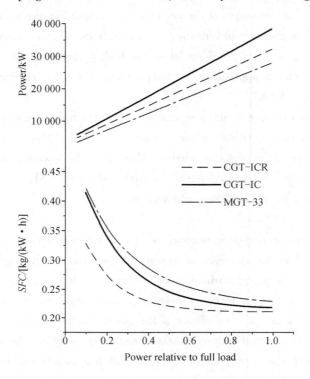

**Fig. 8  Power output and SFC performance**

IC cycle has the highest power output. The use of simple intercooling technology can avoid the piping pressure losses from high pressure compressor outlet to combustor by way of a recuperator, and also the piping pressure losses from power turbine outlet to the atmosphere by way of a recuperator. These losses will arise due to the presence of the recuperator. Because of the above circumstances the IC cycle occupies a more advantaged position in enhancing gas turbine power output as compared with ICR cycle. The reduction of pressure losses can to a certain extent compensate for the loss in cycle efficiency due to the absence of recuperation.

ICR cycle has the highest efficiency. The use of recuperation and inlet guide vane variable geometry technology for the power turbine makes it possible for the ICR to have a relatively high efficiency not only at the design point but also when operating at partial loads.

## Conclusions

1) As viewed from the analysis of ordinary cycles, the use of an intercooling cycle can significantly increase the specific work of gas turbines, but in respect of cycle efficiency the adoption of intercooling may usually bring about a slightly negative influence.

2) An analysis of the IC cycle modification scheme implemented on a specific engine indicates that it is possible to simultaneously enhance its power output and efficiency. This is because there exist the following factors available for taking advantage of.

(1) The physical speed of the high pressure rotor decreases significantly. It is possible to properly adjust its specific speed. Its centrifugal stress also markedly decreases.

(2) Owing to the significant decrease of the air temperature at the compressor outlet, under the precondition of ensuring the blade metal surface temperature of the high temperature turbine not higher than the stipulated value, there is still room for realizing an increase in the turbine inlet temperature.

(3) Due to the decrease of power consumption in the gas generator, a greater portion of energy transfer is completed in the power turbine portion, which has a higher efficiency.

(4) In the renewed matching process of the gas turbine the parameters of some components, such as efficiency, pressure ratio, expansion ratio and temperature etc will change in a more favorable direction.

The results of an analysis indicate that the power output has increased from 28 500 kW to 38 200 kW ($\Delta Ne = 34\%$) and the theremal efficiency from 37% to 38.7% ($\Delta \eta_e = 4.6\%$). The effectiveness of engine overall performance enhancement is significant.

3. The main merits of the CGT-IC are:

1) The changes made to the core component structure of the prototype engine gas generator have been kept to a minimum with the flow path characterized by its maximum versatility. As a result, the high reliability of the prototype engine can be inherited in its entirety.

2) Technical difficult points concerning the reliability of such components as recuperator and power turbine adjustable guide vanes have been sidestepped, resulting in a relatively simple system. All these factors can contribute to shortening development cycle, decreasing development expenses and also enhancing the overall performance of the engine.

3) The performance of the CGT-IC occupies an intermediate position between ICR and a simple cycle. For a developing country which lacks a fully developed aero industry, the scheme described above may be regarded as a reliable and realistic compromise one.

## REFERENCES

[1]   CRISALLI A J, PARKER M L. Overview of the WR-21 intercooled recuperated gas turbine engine system: a modern engine for a modern fleet[C]//Turbo Expo: Power for Land, Sea, and Air. American Society of Mechanical Engineers, 1993, 78903: V03AT15A082.

# Experimental Study on Influence of Cooling Air Flow upon Surface Temperature of High-temperature Turbine Blades in Gas turbine*

WEN Xueyou, XIAO Dongming, GU Zhongming, DU Shugui

(Harbin Marine Boiler & Turbine Research Institute, P. R. China)

**ABSTRACT**

On the basis of the issues occurring in practical production, an experimental study on the influence of cooling media flow upon the surface temperatures of LP turbine blades is carried out on a turbine. The experiment results show that, when the tolerance range of the cooling media flow is narrowed, the surface temperature difference of the blades in blade row is also narrowed evidently, while it has no obvious relation to the "corresponding probability" between the cooling media flow and the blade surface temperature. A few blades even show a "full non-corresponding" relationship. The reason of this phenomenon is that the vortex-matrix type inner cooling structure of blades makes the casting process more complex. The experiment results are helpful to the confirmation of the water flow tolerance in reason and the perfection of the process detail.

**Preface**

The authors have carried out a practical measurement for the surface temperatures of HP and LP turbine blades on a type of marine/industrial gas turbine[1], and then made a comparison and verification between the measured results with the calculational results[2]. The paper, basing on the issues occurring in practical production, develops an experimental study on the influence of cooling media flow used for high temperature turbine blades upon the surface temperatures of blades. In brief, in the case of the theoretical calculation, when the flow rate of the cooling air through a blade is more than the designed value, it is certain that the surface temperature of the blade is less than the calculated value, and vice versa. However, for a blade row of practical precision casting turbine blades, which are designed with a complex vortex-matrix cooling structure, is the conclusion still so simple?

**Experiment equipments introduction**

The measuring object is a row of LP turbine blades on a three-shaft gas turbine (Fig. 1). The blades are designed with a vortex-matrix cooling structure (Fig. 2), which has a preferable cooling

---

\* 本文为2012年美国机械工程师协会涡轮技术会议和博览会论文。

efficiency. Before assembling onto the disk, each one of these LP turbine blades undergoes a water flow experiment on a special water flow test-bed, although trubine blade cooling medium is air(Use air or water flow test result is the stand model. Both can be conversion). The experiment flow is equivalent approximately to the cooling air flow through a blade under the rated condition of the engine. The blades which match the predefined tolerance range are chosen to be assembled. The blades in blade row are numbered and corresponded with the results of the water flow experiment.

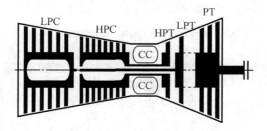

Fig. 1  Schematic of a three-shaft gas turbine

Fig. 2  Structure of cooling air flow channel in LP turbine blades

A Billet – 6 optical pyrometer measuring system is applied(Fig. 3). The temperature sensor receives the heat radiation, which has special effective wavelength, from the surfaces of measured gas turbine blades, then converts this heat radiation into an electrical signal. By processing of this signal, the temperature of the blade located in the observing field of sensor can be obtained; the indicating sensors are assembled on LP rotors, causing the one-to-one correspondence between the measured results and the blade numbers. The measured results are illustrated in the forms of table, histogram and waveform(Fig. 4).

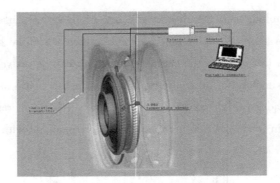

**Fig. 3  Optical pyrometer measuring system**

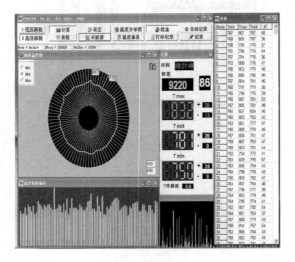

**Fig. 4  Schematic of measured results**

The measured values noticed most in this experiment are listed as below.

$G_{max}$: Maximum value of water flows among blades of this row;

$G_{min}$: Minimum value of water flows among blades of this row;

$G_{mid}$: Average value of water flows on each blade of this row;

$T_{maxmax}$: Maximum value of maximum surface temperatures on each blade of this row;

$T_{maxmin}$: Minimum value of maximum surface temperatures on each blade of this row;

$T_{maxmid}$: Average value of maximum surface temperatures on each blade of this row;

$d_{max}: T_{maxmax} - T_{maxmin}$;

$T_{midmax}$: Maximum value of average surface temperatures on each blade of this row;

$T_{midmin}$: Minimum vale of average surface temperatures on each blade of this row;

$T_{midmid}$: Average value of average surface temperatures on each blade of this row;

$d_{mid}: T_{midmax} - T_{midmin}$;

$T_{minmax}$: Maximum value of minimum surface temperatures on each blade of this row;

$T_{minmin}$: Minimum value of minimum surface temperatures on each blade of this row;

$T_{minmid}$: Average value of minimum surface temperatures on each blade of this row;

$d_{min}$: $T_{minmax} - T_{midmin}$.

It should be said that, because the optical pyrometer is a stationary type, it can scan a strip region on the back of each blade during the measurement (Fig. 5), so that the measured data are all the surface temperatures of blades in the observing field of sensor.

Fig. 5 **Schematic of sensor observing field**

## Experimental study

On No. A gas turbine, with the water flow experiment configuration of $(5\ 300 \pm 650)$ g/min, the surface temperatures of LP turbine blades are measured under the rated condition of gas turbine. The experiment results are reorganized into a relationship chart.

Fig. 6 is the relationship chart of the blade water flows versus the measured average surface temperatures. For the convenience of analyzing, the blades in of this row are sequenced from low value to high value of water flow (black column). The measured average temperature of each corresponding blade is arranged closely to it (gray column). The coordinate on left side is the experimental value of water flow, which is divided into 4 areas from the bottom to the top, marked as $A1$, $A2$, $A3$ and $A4$. The coordinate on right side is the measured average value of blade surface temperature, which is divided into 4 areas from the top to the bottom, marked as $B1$, $B2$, $B3$ and $B4$. The dividing principles are listed as below:

· $G_{mid}$, the average value of water flows on each blade of this row, is regarded as the dividing line of water flow division (the dividing line of area $A2$ and $A3$). This line should be aligned with $T_{midmid}$, the average value of average surface temperatures on each blade of this row (the dividing line of area $B2$ and $B3$);

· $G_{max}$, the maximum value of water flows on blade of this row, is regarded as the upper limit of water flow division (the upper limit of area $A4$). This limit should be aligned with $T_{midmax}$, the Maximum value of average surface temperatures on each blade of this row (the upper limit of area $B1$);

· $G_{min}$, the minimum value of water flows on blade of this row, is regarded as the lower limit of water flow division (the lower limit of area $A1$). This limit should be aligned with $T_{midmin}$ the minimum value of average surface tempeatures on each blade of this row (the lower limit of area $B4$);

· The dividing line between area $A1$ and $A2$ is $(G_{min} + G_{mid})/2$, and should be aligned with the dividing line between area $B_3$ and $B_4$, $(T_{midmin} + T_{midmid})/2$;

· The dividing line between area $A3$ and $A$ is $(G_{max} + G_{mid})/2$, and should be aligned with the dividing line between area $B1$ and $B2$, $(T_{midmax} + T_{midmid})/2$.

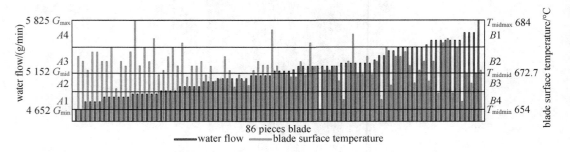

Fig. 6  Relationship chart of blade water flows vs. measured average surface temperatures

In logic, when the water flow lies in area of $A1$ (less), the blade surface temperature should lie in area of $B1$ (high). Namely, when the group levels of area $A$ and area $B$ are same (for example, $A1$-$B1$, $A3$-$B3$, etc.), the logic relationship between both values should be "full corresponding". When the difference of the group levels of area $A$ and area $B$ is 1 (for example, $A1$-$B2$, $A3$-$B2$, $A3$-$B4$, etc.), the logic relationship should be "corresponding". When the difference of the group levels of area $A$ and area $B$ is 2 (for example, $A1$-$B3$, $A2$-$B4$, etc.), the logic relationship should be "non-corresponding". When the difference of the group levels of area $A$ and area $B$ is 3 (for example, $A1$-$B4$, $A4$-$B1$, etc.), the logic relationship should be "full non-corresponding".

Fig. 7 is the relationship chart of blade water flows versus measured maximum surface temperatures.

Fig. 8 is the relationship chart of blade water flows versus measured minimum surface temperatures.

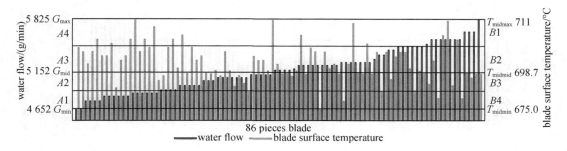

Fig. 7 relationship chart of blade water flows vs. measured maximum surface temperatures

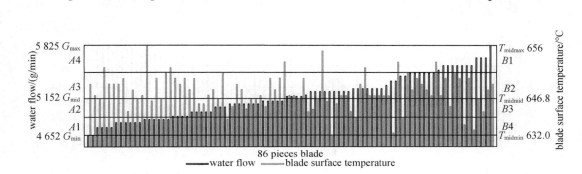

Fig. 8 relationship chart of blade water flows vs. measured minimum surface temperatures

The above experiment results are reorganized into Table 1.

Being afraid of the representative of such one experiment, another experiment is carried out on No. B gas turbine. With the water flow experiment configuration of $(5\,300 - 450) \sim (5\,300 + 400)$ g/min, the surface temperatures of LP turbine blades are measured under the rated condition of gas turbine. The experiment results are showed in Fig. 9 – Fig. 11. The experiment results are reorganized and summarized into Table 2.

Table 1  Summary of measured results of LP turbine blades in No. A gas turbine

| Test item | Full corresponding | Corresponding | Corresponding probability | Non-corresponding | Full non-corresponding | Non-corresponding probability |
|---|---|---|---|---|---|---|
| Average value of blade surface temperature | 30.2% | 58.1% | 88.4% | 9.3% | 2.3% | 11.6% |
| Maximum value of blade surface temperature | 32.6% | 53.5% | 86.0% | 10.5% | 3.5% | 14.0% |
| Minimum value of blade surface temperature | 25.6% | 53.5% | 79.1% | 20.9% | 0.0% | 20.9% |

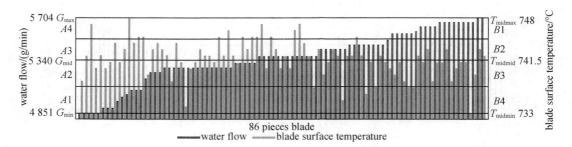

**Fig. 9** Relationship chart of blade water flows vs. measured average surface temperatures

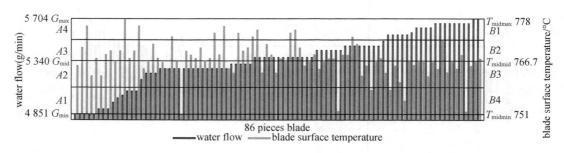

**Fig. 10** Relationship chart of blade water flows vs. measured maximum surface temperatures

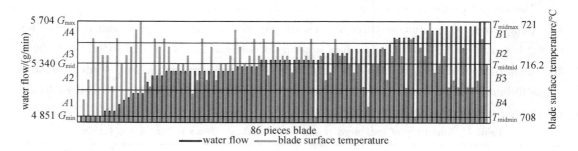

**Fig. 11** Relationship chart of blade water flows vs. measured minimum surface temperatures

Table 2 Summary of measured results of LP turbie blades in No. B gas turbine

| Test item | Full corresponding | corresponding | Corresponding probability | Non-corresponding | Full non-corresponding | Non-corresponding probability |
|---|---|---|---|---|---|---|
| Average value of blade surface temperature | 30.2% | 51.2% | 81.4% | 18.6% | 0.0% | 18.6% |

**Table 2** continued

| Test item | Full corresponding | corresponding | Corresponding probability | Non-corresponding | Full non-corresponding | Non-corresponding probability |
|---|---|---|---|---|---|---|
| Maximum value of blade surface temperature | 31.4% | 54.7% | 86.0% | 14.0% | 0.0% | 14.0% |
| Minimum value of blade surface temperature | 27.9% | 50.0% | 77.9% | 16.3% | 5.8% | 22.1% |

## Analysis of experiment results

1. Comparing the measured results from gas turbines of No. A and No. B (Table 3), it can be found that, since the tolerance range of media flow through LP turbine blade on No. B gas turbine is narrowed, the surface temperature difference of blades in this row is narrowed evidently.

**Table 3** Comparision of measured results from No. A gas turbine and No. B gas turbine

| Turbine | $d_{mid}$/℃ | $d_{max}$/℃ | $d_{min}$/℃ | Tolerance of water flow/(g/min) |
|---|---|---|---|---|
| No. A gas turbine | 30 | 36 | 24 | $G_W \pm 650$ |
| No. B gas turbine | 15 | 27 | 13 | $(G_W - 450) - (G_W + 400)$ |

2. From the measured results summarized in Table 1 and Table 2, it can be found that the proportion of blades with "Full corresponding" and "Corresponding" relationship between water flow and blade surface temperature is about 85%, while the one with "Non-corresponding" relationship is about 15%.

3. Comparing the measured results in Table 1 and Table 2, it can be found that, when the tolerance range of the water flow is narrowed, it has no obvious relation to the "Corresponding probability" between the water flow and the blade surface temperature. The tolerance of water flow through LP turbine blades on No. A gas turbine is more and the "Corresponding probability" is 88%. The tolerance of the water flow through LP turbine blades on No. B gas turbine is less and the "Corresponding probability" is 81%.

4. It should be noticed that, for a few blades with "Full non-corresponding" relationship (0 - 2 blades per stage), their cooling flow is more while the blade surface temperature is higher relatively, or their cooling flow is less while the blade surface temperature is lower relatively. The reason of this phenomenon is complex. There is a need to further improve the manufacturing precision of the inner cooling passage, such as the handwork modification to the wax molding, ect., may cause an excess change of local surface temperature on blade, especially to the tail edge area. The calculation analysis has showed this[2].

5. On the basis of the experiment results, combing with calculational analysis, the deviation of the water flow can be made more perfection.

**REFERENCES**

[1] WEN Xueyou, XUN Baiqiu. Experimental study on surface temperature measurement of high-temperature turbine blades in gas turbine, 9th ISAIF[J]. Korea, 2009.

[2] WEN Xueyou, ZHANG Lichao, GU Zhongming, et al. The validation between the measurement and the calculational analysis for the surface temperature of high-temperature turbine blades in a gas turbine[R]. IGTC 2011-ABS-0019, Japan, 2011.

# An Adjustment Method of Axial Force on Marine Multi-shaft Gas Turbine Rotor[*]

WEN Xueyou, ZHANG Lichao, KOU Dan, XIAO Dongming

(Harbin Marine Boiler and Turbine Research Institute)

ZHANG Han

(China Ship Research and Development Academy)

## ABSTRACT

During the research and development of marine multi-shaft gas turbine, the axial force on the thrust ball bearing of the rotor is required to be measured under actual operating conditions, and then the axial force would be adjusted to a predetermined range with a predetermined method, so as to assure the bearing's long-term and reliable operation.

Basing on the theoretical analysis and engineering practice, we propose a practical method for the analysis and adjustment of the axial force. The method features combining the analysis of the air system and the small deviation analysis for the gas turbine engineering, with the typical experimental verifications.

According to this method and the first measurement results of the axial force, the adjustment measures, which would locate the axial force value in a predetermined range, can be given. Since the proposed adjustment measures are based on the theoretical analysis, the component tests and the actual measurement for the whole engine, the result is of engineering accuracy.

The engine delivery tests also show that the adjusting workload of axial force can be reduced significantly by use of the above method.

## NOMENCLATURE

*TIT*  Turbine inlet temperature
HPT  High pressure turbine
PT  Power turbine
*SFC*  Specific fuel consumption
*Ne*  Power
$n_3$  Speed of power turbine

## Preface

For the gas turbine, the bearing is a key component and has many life-affecting factors. Among them, the axial load is an important one which affects the life of the thrust ball bearing. When the axial force applied on the bearing is excessive or the bearing works under the condition of high speed and light load, the reliability of the bearing would be reduced.

---

[*] 本文为2014年美国机械工程师协会涡轮技术会议与博览会论文。

For either the new-developed engines or the batch products, various factors occurred in the designing, manufacturing and assembling phase may lead the axial force on the gas turbine rotor to be out of the predetermined range. During the research and manufacture of many types of gas turbine, we develop a method which can be used to analyze and adjust the axial force on the rotor of the gas turbine which features with single-shaft, dual-shaft or three-shaft structure. The method can be used in the phases of design, testing and batch production of gas turbine, so as to solve the adjusting problems of the rotor axial force.

In the design phase of a gas turbine, we need to consider the method for adjusting the axial forces on the gas generator rotor and the power turbine rotor running in the whole operating map, as well as set necessary balancing cavity and adjustable sealing structure. In the phase of small batch production, due to some immature factors from manufacturing, assembling, etc., the size of flow path may be out of tolerance, so that the parameter matching among the components of the engine is influenced and the rotor axial force of the engine exceeds the standard. Even the direction of the axial force may be reversed during the process from the low load to the high load. So, making some appropriate adjustments to the area of flow, the clearance of air seal, the aperture, the pressure of chamber, etc., is necessary.

## Design on air system of gas turbine[1]

The main function of the air system is to extract the air with required pressure, temperature and flow rate from the compressor. Making use of the flow path system connected together by internal or external pipelines, pre-swirl nozzles, holes, cavities, etc., the air system can conduct the effects of cooling, sealing, rotor axial force balancing, and so on. See Fig. 1. By adjusting the key size of the air system, we can control the cooling air quantity of extraction and distribution.

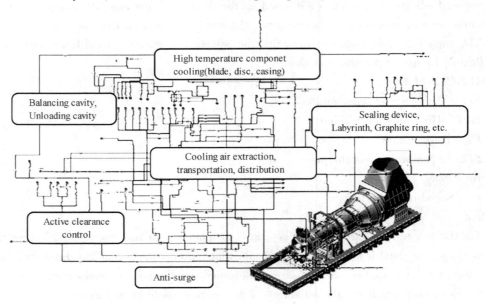

**Fig. 1 Flow chart of air system in gas turbine**

Any air system from a gas turbine can be abstracted as a flowing network composed of pressure sources, throttle units and chambers. For an air system containing throttle units of $m$ and chambers of $n$, its temperature, pressure and flow can be described with $m+n$ dimensional nonlinear equation set which is composed of the momentum equation, the continuity equation, and the energy equation. i. e., the mathematical model of air system is a nonlinear equation set. Its general expression is as follows:

$$f_i(X) = 0 \quad i = 1,2,3,\cdots,(m+n) \tag{1}$$

In equation set, $X$ is a vector(one-dimensional array) made up of unknown flow($m$), temperature ($T$) and pressure($P$)

We make use of the Flowmaster V7 (Gas turbine) to conduct the system equilibrium calculation. Based on the air flow path structure of the gas turbine, the one-dimensional calculation model of the air system is established, as shown in Fig. 1. All kinds of structures of sealing, rotary chamber, inlet pre-swirl, etc., are taken into account in the air system. By use of the equilibrium calculation, the pressure and flow of each chamber in the gas turbine can be obtained. The calculation precision can also be validated by measuring the pressures in some chambers. Then, the system model may be modified so as to approach the real state closer.

**Analysis on rotor axial force**

The rotor axial force is mainly a resultant force of the three parts, i. e., the axial force on the flow path through the compressor, turbine and the axial force on the internal chambers of the gas turbine.

In the design process of gas turbine, in addition to the analysis on the axial force under the rated power, the axial load under the partial load should also be analysed, so that the changing trend of the rotor axial load to the power can be obtained. Due to the different structure and characteristics of each engine type, the changing laws of the axial forces to the power are not the same.

By means of the calculation and analysis, under the design conditions and some part-load conditions, the axial forces with designed aerodynamic parameters, clearances and throttle diameters, can be obtained. However, under the actual situation, when some factors having great influence on the axial force(such as the diameter of the rear balancing disc and the clearance of the rear seal in HP compressor, and the throat area of the turbine nozzle) have some change, the axial force would change a lot.

The design goal is that the axial force can be located in the specific range during the whole actual operating condition, i. e., the maximum axial force is lower than the specified upper limit, the minimum axial force is higher than the specified lower limit, and the direction of axial force remains unchanged. When the axial force exceeds the allowable limits, the necessary adjustments of the local structure, the clearance, the pressure of the balancing cavity, etc., is needed.

Therefore, combining the calculation and analysis of the air system with the small deviation analysis method for gas turbine engineering, we develop a method used to analyze and adjust the axial force.

**Application of small deviation method in gas turbine axial force adjustment**

To give the modelling and the algorithm of the operation of a gas turbine is complex, because a complex nonlinear mathematical model is needed to describe the procedure. In this paper, we only discuss the "small deviation" between the design values and the actual ones. In the engineering, the approximate linearization method is used usually for these cases.

The small deviation method is a method to linearize the relation of some phenomenon. So, when the operation procedure of an engine is transformed into a small deviation equation, the number of the equations or unknowns will not be changed. The solvability of the equations is same with the original equations.

According to the characteristics of marine gas turbine, the flow state through the turbine guide blade throat is not necessarily close to the critical. Then the assumption of "critical nozzle" occurring in the small deviation method of aero-engine can be thrown away and the calculation precision can be enhanced.

By use of the small deviation equation of each turbine stage, the correlation matrix, including the independent variables such as the throat area of each turbine stage, as well as the engine dependent variables, can be obtained (See Expression 2)

$$AX = BY \tag{2}$$

In the expression $X = \{x_1, x_2, x_3, \cdots, x_n\}$, here $X$ is the matrix of the engine dependent variables. For a three-shaft gas turbine, the dependent variables, $X_i$, includes the power, the specific fuel consumption, the speed, the air flow rate, the gas flow rate, the gas temperature, the pressure ratio of each compressor, the expansion ratio of each blade row of turbine, etc. (total 32 items).

$Y = \{y_1, y_2, \cdots, y_m\}$, here $Y$ is the matrix of the engine independent variables. For a three-shaft gas turbine, the independent variables, $Y_i$, includes the turbine inlet temperature, the total pressure recovery coefficient of each portion, the efficiency of each component, the quantity of the cooling air extraction, the blade throat area of each turbine stage, etc. (total 28 items).

$A = (a_{ij})(i, j = 1, 2, 3, \cdots, n)$, here $A$ is the coefficient matrix of the characteristic parameters for the engine independent variables of small deviation.

$B = (b_{ik})(i = 1, 2, 3, \cdots, n)(k = 1, 2, 3, \cdots, m)$, here $B$ is the coefficient matrix of the characteristic parameters for the engine dependent variables of small deviation.

Now, we would calculate the rotor axial force. The expression is as follows:

$$F = \sum_{i=1}^{n}(p_i \cdot f_i + \Delta \nu_i \cdot m/g) \tag{3}$$

In the research on axial force adjustement of gas turbine in this paper, the objects include the single-shaft gas turbines, the dual-shaft gas turbines and the three-shaft gas turbines. The number of their HP turbine stage, LP turbine stage and power turbine stage may be random. We establish the matrix of the small deviations, with the each blade row throat area of turbine stage as an independent variable. Then, we obtain the influence coefficient of the throat area change in each blade row of turbine stage to each engine dependent variable. Therefore, we can analyse the share of the throat area tolerance, due to the manufacturing and the assembling of the turbine blade in corresponding stage, among the tolerance of the axial force. By overall consideration of all these factors, the adjustment measures can be determined.

Fig. 2 shows the full process including the calculation, the example and the adjustment of the axial force in gas turbine.

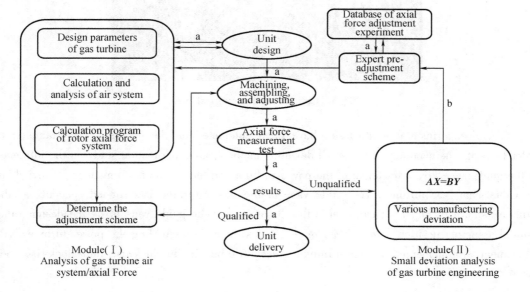

Fig. 2 Flow chart

Firstly, following the a instruction, the gas turbine design department accesses the module (Ⅰ), the "Analysis modules of gas turbine air system/axial force". If the measuring result of the axial force doesn't meet the requirement, the design department should access the module (Ⅱ), the "Small deviation analysis module of gas turbine engineering", following the b instruction. Then, the module (Ⅱ) would output the adjustment suggestion. After being adjudged by experts, the suggestion would be input into the module (Ⅰ) and be implemented.

## Verification test by single-factor

In order to ensure that the above analysis software can meet the engineering accuracy, some single-factor verification tests are carried out. These factors include all kinds of seal devices. They can be tested on the sealing test-beds so that the effect of these single-factors (clearance, pressure difference) can be obtained. See Fig. 3.

Fig. 3　Comprehensive seal test-bed

Several experimental verifications have been conducted on the engines. For example, in order to verify the accuracy of the small deviation method, an experimental study on the influence of the guide vane area adjustment of the power turbine, on the engine performance, is carried out on a dual-shaft gas turbine. The guide vane of the power turbine is designed specially as the structure shown in Fig. 4. It can enable the throat area of the guide vane to be the design value (initial assembly position), or $+2\%$, or $-2\%$, without disassembling the power turbine. The adjustment is implemented through turning the eccentric bushing in clockwise/anti-clockwise. See Fig. 4.

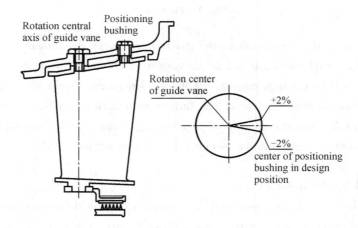

Fig. 4　The controlling mechanism of the guide vane throat area

The test is undertaken in three guide vane conditions, including the rated position, the increase of $2\%$, and the decrease of $2\%$ throat area. Under each condition, a measurement in the power range of $35\% - 100\%$ is carried out. Table 1 shows the influence of the throat area change in the power turbine guide vane on the gas turbine parameters under rated power.

**Table 1  The table of influence coefficient with power turbine nozzle area change of +1%**  %

| Data source | Test result | Calculation results of small deviation method |
|---|---|---|
| Pressure ration of compressor | -0.48 | -0.31 |
| TIT | -1.05 | -0.85 |
| Expansion ratio of HPT | 0.59 | 0.54 |
| Expansion ratio of PT | -1.01 | -0.84 |
| Ne | -2.89 | -2.22 |
| SFC | 0.83 | 1.08 |

## Measuring and adjusting of axial force on gas generator[4]

On the gas generator of a three-shaft gas turbine, a test is carried out concerning the influence of the multi-factor adjustment on the axial force of the high pressure rotor. The results of the first testing are as follows: under the operating condition at 35% of rated power, the axial force of the high pressure rotor reaches the highest value, 24.31kN; under the operating condition of full power, the axial force falls to 9.69 kN. It shows that the axial force exceeds the permitted range greatly under low operation condition. By applying the software analysis described in this paper, we quickly put forward following adjustment schemes.

1. According to the recommended value, replace the rear seal disk of the HP compressor with a smaller one in diameter, the rear air seal disk has a variety of different diameters (See Fig. 5);

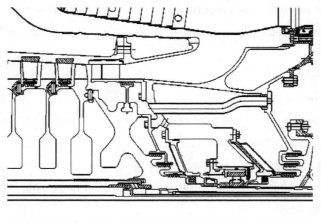

Fig. 5  Schematic diagram of rear seal in HP compressor

2. According to the recommended value, adjust the rear seal clearance of HP compressor;

3. According to the recommended value, adjust the sealing clearance between the HP turbine rotor blade and the LP turbine guide vane.

After adjusting according to the scheme, from the test date obtained, it is shown that the maximum axial force of the gas generator rotor is 14.85 kN at the condition of 20% power, and the axial force at the condition of full power is 5.02 kN, which meets the design requirements well (See Fig. 6).

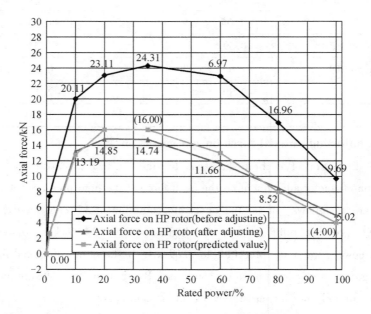

Fig. 6　Experimental verification curve of axial force on gas generator

## Measuring and adjusting of axial force on power turbine rotor[4]

The bearing structure of the power turbine rotor in marine gas turbine has various practical forms, including:

1. With rolling bearing;
2. With sliding bearing;
3. With combination bearing (the front end uses the roller bearing, the rear end uses the ball bearing and the tilting-pad thrust bearing).

When the rotor is equipped with a sliding bearing structure or a combination bearing structure, because the tilting-pad thrust bearing can withstand more axial load, the adjustment of the rotor axial force is simple relatively.

**Example**

For a set of marine gas turbine, which is derived from an aero turbojet engine and is equipped with a rolling bearing structure in its power turbine (see Fig. 7), we have designed a set of device for measuring and automatic controlling of the axial force of power turbine rotor.

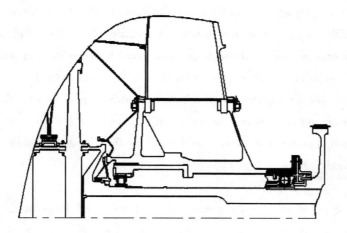

**Fig. 7  Power turbine section**

Standard strain force-measuring ring is adopted for the measurement of the rotor axial force.

For the gas turbine used for ship propulsion, we measure in detail the axial force of each point in the range of engine-propeller (including fixed pitch propeller and variable pitch propeller) matching characteristics (See Fig. 8). By adjusting the opening degree of the electric plug under each specified operating condition, we can control the balancing cavity pressure on the rear side of the power turbine disc, and the axial force can be kept as a fixed value. Then, we can record the corresponding plug opening, the balancing chamber pressure, the speeds of gas generator and power turbine, the front and rear pressures of power turbine disc, the sensor temperatures, and all operating parameters of the engine.

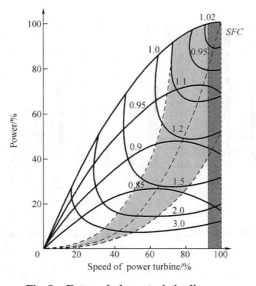

**Fig. 8  External characteristic diagram**

According to the study analysis on the measured data, we determine ultimately that, by using an orifice plate with specific diameter instead of electric plug, the axial force still can be controlled in a reasonable range, although the difference between the maximum value and the minimum value of the axial forces on some point increases to some extent.

Considering that the gas turbine can also be used to drive a generator for the cogeneration, or to drive a compressor unit for the natural gas boosting, we have also measured the axial forces under constant speed operating conditions and determined the diameters of their equivalent orifices by use of this method.

**Conclusion**

The method proposed for analysing and adjusting the axial force in this paper has been applied in new gas turbine prototypes and batch production engines for several years.

Among these engines, one set has undergone a 3 000-hour endurance test. Through the disassembly examination, it is found that the bearing in it is still in good condition.

According to the experiences of the actual measurement, analysis and adjustment, aiming at the pratical model, we compiled a "Working instruction of axial force adjustment". By use of this instruction, the working efficiency of the axial force adjustment is enhanced significantly.

Fig. 9 shows the measurement and adjustment of the axial forces on the first produced 12 sets of one three-shaft marine gas turbine type. The vertical coordinate represents the measuring times (black colour) and the adjusting times (gray colour) of the axial forces. The method described in this paper has not been applied completely to the adjustment of the initial 4 sets. By adopting the method, the adjusting times of the axial forces are reduced significantly.

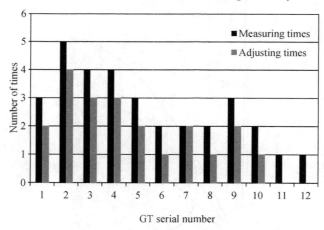

**Fig. 9 Statistical chart of measured axial force on gas turbine**

When the overrun of the axial force measuring result beyond the predetermined value is small, we can make corresponding adjustment according to the "Working instruction of axial force adjustment" rather than making additional axial force tests. This adjustment is not included into the adjusting times of the axial force.

**REFERENCES**

[1] WANG Gehua. Aero Engine Design Manual: air system and heat transfer analysis, Volume 16 [M]. Beijing: Aviation Industry Press, 2001.

[2] CHERKEZ. The small deviation analysis of gas turbine engineering. [In Russia], 1965.

[3] WEN Xueyou. Gas turbine performance analysis[R]. Resarch Report 1998-03-11. (Inside)

[4] ZHANG Lichao, PAN Hongwei, WEN Xueyou. General report on measurement and adjustment of axial force of marine three-shaft gas turbine[R]. Research report 2005. (Inside)

# An Evaluation of the Application of Nanofluids in Intercooled Cycle Marine Gas Turbine Intercooler

ZHAO Ningbo
(College of Power and Energy Engineering Harbin Engineering University)
WEN Xueyou
(Harbin Marine Boiler & Turbine Research Institute)
LI Shuying
(College of Power and Energy Engineering Harbin Engineering University)

## ABSTRACT

Coolant is one of the important factors affecting the overall performance of the intercooler for the intercooled cycle marine gas turbine. Conventional coolants such as water and ethylene glycol have lower thermal conductivity which can hinder the development of highly effective compact intercooler. Nanofluids that consist of nanoparticles and base fluids have superior properties like extensively higher thermal conductivity and heat transfer performance compared to those of base fluids. This paper focuses on the application of two different water-based nanofluids containing aluminum oxide ($Al_2O_3$) and copper (Cu) nanoparticles in intercooled cycle marine gas turbine intercooler. The effectiveness-number of transfer unit method is used to evaluate the flow and hear transfer performance of intercooler and the thermophysical properties of nanofluids are obtained from literature. Then the effects of some important parameters such as nanoparticle volume concentration, coolant Reynolds number, coolant inlet temperature and gas side operating parameters on the flow and heat transfer performance of intercooler are discussed in detail. The results demonstrate that nanofluids have excellent heat transfer performance and need lower pumping power in comparison with base fluids under different gas turbine operating conditions. Under the same heat transfer, Cu-water nanofluids can reduce more pumping power than $Al_2O_3$-water nanofluids. It is also concluded that the overall performance of intercooler can be enhanced when increasing the nanoparticle volume concentration and coolant Reynolds number and decreasing the coolant inlet temperature.

## NOMENCLATURE

$A$    Heat transfer surface area, $m^2$

$A_{ff}$    Free flow area, $m^2$

$C$    Heat capacity rate, W/K

$Cr$    $C_{min}/C_{max}$

---

\* 本文为2015年美国机械工程师学会涡轮技术会议与展览会论文。

$C_p$  Specific heat, J/kg·K
$D$  Hydraulic diameter, m
$f$  Fanning factor
$G$  Mass flow velocity, kg/(m²·s)
$h$  Convective heat transfer coefficient, W/(m²·K)
$H$  Fin height, m
$j$  Colburn factor
$L1$  Intercooler length, m
$L2$  Intercooler width, m
$L3$  Intercooler height, m
$m$  Fin parameter factor
$N$  Number of fin layers
$NTU$  Number of transfer units
$Nu$  Nusselt number
$P$  Pressure drop, Pa
$PP$  Pumping power, W
$Pr$  Prandtl number
$Q$  Heat transfer rate, W
$Re$  Reynolds number
$S_f$  Fin pitch, m
$T$  Temperature, K
$t_f$  Fin thickness, m
$U$  Overall heat transfer coefficient, W/(m²·K)
$V$  Volumetric flow rate, m³/s
$W$  Mass flow rate, kg/s

**Greek letters**

$\beta$  The aspect ratio of channel
$\eta_{ef}$  Effective heat transfer surface efficiency
$\eta_f$  Heat transfer efficiency of fin
$\delta$  Thickness, m
$\lambda$  Thermal conductivity, W/(m·K)
$\mu$  Viscosity, Pa·s
$\rho$  Density, kg/m³
$\varepsilon$  Effectiveness
$\phi$  Nanoparticle volume concentration

**Subscripts**

    $a$    gas
    $b$    coolant
    bf   base fluid
    nf   nanofluids
    $p$    nanoparticle
    pl    plate
    s     seal
    sp   side plate

**Preface**

    Intercooled (IC) cycle technology is a feasible method to develop high power marine gas turbines[1]. As an important part of intercooled cycle marine gas turbine, the intercooled system including intercooler and sea water heat exchanger is used to decrease the temperature of gas from the low pressure compressor in order to reduce the power consumption of high pressure compressor and increase the output power of the whole marine gas turbine. Compared with the sea water heat exchanger, the intercooler is more important for the intercooled system because it can directly influence the structure and perfomance of an engine. Therefore, how to realize the optimization design of high efficiency compact intercooler is one of the important problems for intercooled cycle marine gas turbine[2].

    Plate-fin heat exchangers are characterized by a high heat transfer surface per unit volume, which can result in a higher efficiency than the conventional heat exchanger. Therefore, it is generally used for intercoolers. With strict requirements for space and efficiency, different kinds of fin surfaces (such as wavy fin, triangle fin, serrated fin and louver fin) are developed to increase the heat transfer area and enhance the convective heat transfer coefficient[3]. However, these improvements have already reached their limitation and many of above approaches cannot be directly used in intercoolers because the pressure drop must be considered when enhancing the heat transfer perfomance of the intercooler. If the pressure drop of the gas is too high, it will result in an additional burden on the high pressure compressor to keep a constant pressure ratio, which means more fuel consumption.

    Coolant seems to be another crucial factor in determining the performance of intercooler. Conventional coolants such as water and ethylene glycol have lower thermal conductivity which can hinder the development of a highly effective compact intercooler. In comparison with liquid, many solids like metal and metal oxides have higher thermal conductivity. For example, water has a thermal conductivity of 0.618 W/(m·K) at 303 K while the thermal conductivity of Cu is 401 W/(m·K)[4]. This finding provides an effective method for increasing the thermal conductivity of liquid by suspending solids in liquid. In order to solve the problems of clogging and abrasion, a new solid-liquid mixture consists of nanoparticles and base fluid was developed and named nanofluids by Choi in 1995[5]. In recent years, many investigators have studied the various features of nanofluids by using theoretical and experimental methods[6-12]. Their results showed nanofluids had superior properties like extensively greater thermal conductivity and heat transfer performance as compared to the base fluids. Lee et al.[13] measured the thermal conductivity of copper oxide (CuO)-ethylene glycol nanofluids and observed that the thermal conductivity of nanofluids could be enhanced up to 20% compared to that of base fluid with the adtition of 4% CuO nanoparticles. Xuan and Li[14] presented a study on the thermal conductivity of a nanofluids consisting of Cu nanoparticles and water. Their measured results showed that the ratio of Cu-water nanofluids thermal conductivity and that of base liquid increased from 1.24 to 1.78 when the nanoparticles volume concentration varied from 2.5% to 7.5%. In addition, Xuan and Li[15] also pressented an experimental investigation on the convective heat transfer and flow feature of Cu-water nanofluids. The results indicated that the suspended nanoparticles could enhance the heat transfer of base fluid and the greater heat transfer enhancement was found to be more than 39% at 2% nanoparticle volume concentration. More reserarch can refer to the references[16 – 17].

Considering the superior characteristics of nanofluids, much theoretical research has been performed for evaluating the performance of different heat exchangers (such as automotive radiator, plate heat exchanger and spiral flat tube heat exchanger) by using nanofluids as coolant[18]. Leong et al. investigated the heat transfer characteristics of an automotive car radiator using ehtylene glycol based Cu nanofluids as coolants[19]. Their results showed that the overall heat transfer could be increased with the usage of nanofluids and about 3.8% of heat transfer enhancement could be obtained with the addition of 2% Cu nanoparticle in base fluids. They also observed that nanoparticle volume concentration, air and coolant Reynolds number were the important factors influencing the heat transfer performance of an automotive car radiator when nanofluids were used as coolants. Besides, Ray and Das performed a detailed computational study to compare the fluid dynamic and thermal performance of three nanofluids in an automotive radiator operating in the turbulent regime[20]. They found that the heat transfer could be enhanced by using nanofluids as heat transfer mediums and nanofluids showed a reduction in pumping power over base fluid when the nanoparticle volume concentration is smaller. More similar studies can be found in references[21 – 25].

To the best of our knowledge, little study has been published about the research of intercooled cycle marine gas turbine intercooler using nanofluids as coolant. Therefore, this study investingates the flow and heat transfer performance of intercooled cycle marine gas turbine intercooler operated with nanofluids. As the most commonly used nanofluids, two different water-based nanofluids containing $Al_2O_3$ and Cu nanoparticles are seleted to evaluate the effects of different parameters (such as nanoparticle volume concentration, coolant Reynolds number, coolant inlet temperature and gas side operating parameters) on the flow and heat transfer performance of intercooler.

## Mathematical models of intercooler

Fig. 1 and Fig. 2 depict a schematic view of a reverse flow plate-fin intercooler with straight fins and the basic geometric structure of fins, respectively. The following assumptions will be sued for the analysis.

(1) The straight fin is used on the gas and coolant side.

(2) The number of fin layers for the gas side ($N_a$) is assumed to be one more than the coolant side ($N_b$). That is, $N_a = N_b + 1$.

(3) Intercooler works in a steady state condition.

(4) The thickness of all the fins is assumed uniform and the thermal resistance of fins is negligible.

(5) All the parts are made of same material.

(6) The influences of fouling and corrosion are neglected.

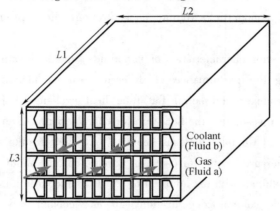

**Fig. 1  Schematic representation of reverse flow plate-fin intercooler**

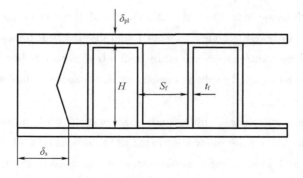

**Fig. 2  Datailed view of straight fin**

In this study, since the output temperature of the fluids is unspecified, the effectiveness-nubmer of transfer unit ($\varepsilon$-$NTU$) method is used to evaluate the flow and heat transfer performance of the intercooler in the modeling process. The effectiveness of reverse flow plate-fin intercooler is proposed as

$$\varepsilon = \frac{1 - \exp[-NTU(1 - Cr)]}{1 - Cr\exp[-NTU(1 - Cr)]} \quad (1)$$

where $Cr = C_{\min}/C_{\max}$ —heat capacity ratio;

$NTU$—the number of transfer units.

Considering the thermal resistance of the plates, $NTU$ can be determined by

$$\frac{1}{NTU} = \frac{C_{\min}}{UA} \quad (2)$$

$$\frac{1}{UA} = \frac{1}{\eta_{ef,a} h_a A_a} + \frac{\delta_{pl}}{\lambda_{pl} A_{pl}} + \frac{1}{\eta_{ef,b} h_b A_b} \quad (3)$$

Considering the turbulent flow of gas, the heat transfer coefficient is calculated in terms of Gnielinsk[26] which is given as

$$h_a = \frac{Nu_a \lambda_a}{D_a} \quad (4)$$

Nusselt number and hydraulic diameter can be calculated as follows:

$$Nu_a = \frac{(f_a/2)(Re_a - 1\,000) Pr_a}{1 + 12.7(f_a/2)^{0.5}(Pr_a^{2/3} - 1)} \quad (5)$$

$$D_a = \frac{2(H_a - t_{f,a})(S_{f,a} - t_{f,a})}{H_a + S_{f,a} - 2t_{f,a}} \quad (6)$$

Reynolds number and Prandtl number of gas are defined as below:

$$Re_a = \frac{W_a D_a}{A_{ff,a} \mu_a} \quad (7)$$

$$Pr_a = \frac{C_{p,a} \mu_a}{\lambda_a} \quad (8)$$

And the Fanning factor $f_a$ is given as

$$f_a = \frac{1}{(1.58 \ln Re_a - 3.28)^2} \quad (9)$$

For the coolant which is in laminar flow, the heat transfer coefficient is calculated in terms of Colburn factor[27] which is given as

$$h_b = \frac{j_b G_b C_{p,b}}{Pr_b^{2/3}} \quad (10)$$

The Colburn factor $j_b$ is given as

$$j_b = \exp[0.103\,109(\ln Re_b)^2 - 1.910\,91\ln Re_b + 3.211] \quad (11)$$

Mass flow velocity of coolant can be obtained as follows:

$$G_b = \frac{W_b}{A_{ff,b}} \quad (12)$$

Reynolds number and Prandtl number of coolant are defined as below:

$$Re_b = \frac{W_b D_b}{A_{\text{ff},b}\mu_b} \tag{13}$$

$$Pr_b = \frac{C_{p,b}\mu_b}{\lambda_b} \tag{14}$$

In this study, the effective circulation area for the two sides are formulated as

$$A_{\text{ff},a} = \frac{N_a(L2 - 2\delta_s)(H_a - t_{\text{f},a})(S_{\text{f},a} - t_{\text{f},a})}{S_{\text{f},a}} \tag{15}$$

$$A_{\text{ff},b} = \frac{N_b(L2 - 2\delta_s)(H_b - t_{\text{f},b})(S_{\text{f},b} - t_{\text{f},b})}{S_{\text{f},b}} \tag{16}$$

The heat transfer areas of intercooler for the two sides are calculated by

$$A_a = 2N_a L1(L2 - 2\delta_s)\left(1 + \frac{(H_a - 2t_{\text{f},a})}{S_{\text{f},a}}\right) \tag{17}$$

$$A_b = 2N_b L1(L2 - 2\delta_s)\left(1 + \frac{(H_b - 2t_{\text{f},b})}{S_{\text{f},b}}\right) \tag{18}$$

Heat transfer efficiencies of heat transfer surface for the two sides are obtained by

$$\eta_{\text{ef},a} = \frac{(S_{\text{f},a} - t_{\text{f},a}) + \eta_{\text{f},a}(H_a - t_{\text{f},a})}{S_{\text{f},a} + H_a - 2t_{\text{f},a}} \tag{19}$$

$$\eta_{\text{ef},b} = \frac{(S_{\text{f},b} - t_{\text{f},b}) + \eta_{\text{f},b}(H_b - t_{\text{f},b})}{S_{\text{f},b} + H_b - 2t_{\text{f},b}} \tag{20}$$

Heat transfer efficiencies of fin for the two sides are formulated as

$$\eta_{\text{f},a} = \frac{\tan(0.5 m_a H_a)}{0.5 m_a H_a} \tag{21}$$

$$\eta_{\text{f},b} = \frac{\tan(0.5 m_b H_b)}{0.5 m_b H_b} \tag{22}$$

Fin factors for the two sides are defined as follows:

$$m_a = \sqrt{\frac{2h_a}{\lambda_{\text{f},a} t_{\text{f},a}}} \tag{23}$$

$$m_b = \sqrt{\frac{2h_b}{\lambda_{\text{f},b} t_{\text{f},b}}} \tag{24}$$

Therefore, the heat transfer rate is obtained as follws:

$$Q = \varepsilon C_{\min}(T_{\text{in},a} - T_{\text{in},b}) \tag{25}$$

To simplify the computation, this study only considers the core part pressure drop of coolant which is defined as below:

$$P_b = \frac{2 f_b L1 G_b^2}{\rho_b D_b} \tag{26}$$

And the Fanning factor $f_b$ is given as

$$f_b = \frac{24}{Re_b}(1 - 1.3553\beta_b + 1.9467\beta_b^2 - 1.7012\beta_b^3 + 0.9564\beta_b^4 - 0.2537\beta_b^5) \tag{27}$$

where, $\beta_b$ is aspect ratio of coolant channel and it is defined as

$$\beta_b = \frac{S_{f,b} - t_{f,b}}{H_b - t_{f,b}} \tag{28}$$

Then, the pumping power can be expressed as

$$PP_b = P_b V_b \tag{29}$$

where, $V_b$ is the volumetric flow rate of coolant and it is obtained by

$$V_b = \frac{W_b}{\rho_b} \tag{30}$$

**Thermophysical Properties**

As previously mentioned, thermophysical properties of fluids are the primary influencing factors of the flow and heat transfer. In order to improve the modeling accuracy, the thermophysical properties dependent on temperature should be considered because the temperature difference between inlet and outlet of the gas or coolant are very big.

For the gas, the thermophysical properties can be curve-fitted by using the data from Yang[28], which presents a broader temperature from 273 K to 773 K. Among them, the density correlation is derived from the ideal gas law. The correlations are given as follows.

**Density**

$$\rho = -1.487 \times 10^{-9}(T-273)^3 + 3.638 \times 10^{-6}(T-273)^2 - 0.003\,088(T-273) + 1.243\,5 \tag{31}$$

**Specific heat**

$$C_p = 4 \times 10^{-4}(T-273)^2 + 0.02(T-273) + 100\,3 \tag{32}$$

**Thermal conductivity**

$$\lambda = 2.456 \times 10^{-4} T^{0.823} \tag{33}$$

**Viscosity**

$$\mu = 1.506\,19 \times 10^{-6} \frac{T^{1.5}}{T+122} \tag{34}$$

Water, as base fluid, is the most common and low cost coolant. The investigation indicated that the thermophysical properties were weakly dependent on the pressure but greatly influenced by temperature. Correlations for the thermophysical properties of water dependent on temperature (273 K $\leq T \leq$ 373 K) are developed in this study based on the reported data from National Institute of Standards and Technology(NIST)[29].

**Density**

$$\rho = 1.976\,72 \times 10^{-5}(T-273)^3 - 6.591\,68 \times 10^{-3}(T-273)^2 + 0.050\,978\,7(T-273) + 1\,000.140\,67 \tag{35}$$

**Specific heat**

$$C_p = -1.568\,41 \times 10^{-4}(T-273)^3 + 0.0381\,57(T-273)^2 - 2.276\,675(T-273) + 4\,217.959\,0 \tag{36}$$

**Thermal conductivity**
$$\lambda = -1.823\ 61 \times 10^{-8}(T-273)^3 - 6.629\ 243 \times 10^{-6}(T-273)^2 + 2.031\ 767 \times 10^{-3}(T-273) + 0.561\ 150 \tag{37}$$

**Viscosity**
$$\mu = -2.591\ 81 \times 10^{-6}(T-273)^3 + 5.889\ 95 \times 10^{-4}(T-273)^2 - 0.048\ 284(T-273) + 1.784\ 89 \tag{38}$$

Over the past decade, a large number of studies have been widely done on the thermophysical properties measurement[30-31] and theoretical model[32-33] for different nanofluids. The available investigations indicated the density and specific heat of nanofluids can be theoretical described based on the physical principle of the mixture rule[34]. However, it is a fact that the existing models still cannot explain the reasons for the enhanced thermal conductivity and viscosity of nanofluids. Although the influences of some parameters such as temperature, nanoparticle volume concentration and nanoparticle size have been discussed in existing studies, the overall physical description was still unavailable. Concerning above problems, the following classical models are used to obtain the thermophysical properties of different nanofluids in this study.

**Density**[35]
$$\rho_{nf} = (1-\phi_p)\rho_{bf} + \phi_p\rho_p \tag{39}$$

**Specific heat**[36]
$$C_{p,nf} = \frac{(1-\phi_p)\rho_{bf}C_{p,bf} + \phi_p\rho_p C_{p,p}}{\rho_{bf}} \tag{40}$$

**Thermal conductivity**[37]
$$\lambda_{nf} = \frac{\lambda_p + (n-1)\lambda_{bf} + (n-1)\phi_p(\lambda_p - \lambda_{bf})}{\lambda_p + (n-1)\lambda_{bf} - \phi_p(\lambda_p - \lambda_{bf})}\lambda_{bf} \tag{41}$$

where, $n$ is shape factor of nanoparticles and $n=3$ when the nanoparticles are sphere.

**Viscosity**[38]
$$\mu_{nf} = \frac{\mu_{bf}}{(1-\phi_p)^{2.5}} \tag{42}$$

## Results and discussions

In this study, Table 1 shows the preliminary structural data of the intercooler obtained from Wen[39] which is used to analyze the flow and heat transfer performance of an intercooler by using nanofluids. It is assumed that there are ten identical intercooler modules in the practical application. Table 2 lists the inlet parameters of the gas under different gas turbine operating conditions for one intercooler module. The intercooler is made of copper-nickel alloy and its thermal conductivity, specific heat and density are 38.5 W/(m·K), 380 J/kg·K and 8 890 kg/m³, respectively, Water is chosen as the base fluid and the properties of $Al_2O_3$ and Cu nanoparticles are listed in Table 3.

Based on the $\varepsilon$-NTU method mentioned above, comparative analysis of water and water-based nanofluids on the flow and heat transfer performance of intercooler with different nanoparticle volume concentration, coolant Reynolds number and inlet temperature and gas side operating paramenters are discussed in detail.

## Table 1 Preliminary structure of the intercooler[39]

| Parameters | Values |
|---|---|
| Number of gas side fin layers, $N_a$ | 36 |
| Fin pitch of gas side, $S_{f,a}$/m | $1.4 \times 10^{-3}$ |
| Fin height of gas side, $H_a$/m | $6.2 \times 10^{-3}$ |
| Fin thickness of gas side, $t_{f,a}$/m | $2 \times 10^{-4}$ |
| Number of coolant side fin layers, $N_b$ | 35 |
| Fin pitch of coolant side, $S_{f,b}$/m | $1.4 \times 10^{-3}$ |
| Fin height of coolant side, $H_b$/m | $3 \times 10^{-3}$ |
| Fin thickenss of coolant side, $t_{f,b}$/m | $2 \times 10^{-4}$ |
| Side plate thickness, $\delta_{sp}$/m | $1.6 \times 10^{-3}$ |
| Plate thickness, $\delta_{pl}$/m | $5 \times 10^{-4}$ |
| Seal thickness, $\delta_s$/m | $6 \times 10^{-3}$ |
| Intercooler length, $L1$/m | 0.35 |
| Intercooler width, $L2$/m | 0.426 6 |

## Table 2 The inlet parameters of gas under different gas turbine operating conditions[39]

| Operating conditions | Inlet pressure/Pa | Inlet temperature/K | Inlet mass flow rate/(kg/s) |
|---|---|---|---|
| 100% | 302 963 | 428.25 | 7.293 |
| 85% | 288 460 | 422.45 | 6.852 |
| 70% | 277 647 | 416.55 | 6.410 |
| 58% | 258 919 | 408.55 | 5.736 |
| 47% | 240 192 | 400.35 | 5.063 |
| 35% | 221 464 | 392.25 | 4.389 |
| 17% | 170 000 | 380.25 | 3.290 |

## Table 3 Thermophysical parameters of nanoparticle materials

| Material type | Thermal conductivity /(W/(m·K)) | Specific heat capacity /(J/(kg·K)) | Density/(kg/m³) |
|---|---|---|---|
| Cu | 401 | 386 | 8 960 |
| $Al_2O_3$ | 36 | 765 | 3 600 |

## Effects of nanoparticle volume concentration

Nanoparticle volume concentration is one of the important factors influencing the thermophysical properties of nanofluids, which can further influence the overall flow and heat transfer performance of an intercooler. Most literatures indicated that the density, thermal conductivity and viscosity of nanofluids increased with the increase of nanoparticles volume concentration. According to the existing experimental and theoretical results, the specific heat of nanofluids decreased when increasing the volume concentration of nanoparticles. As the nanoparticle volume concentration increases, thermal conductivity of nanofluids will increase, which is conducive to enhance the heat transfer of the intercooler. However, the increase of nanofluids viscosity is not favorable because it can lead to increase the pressure drop of nanofluids. In this section, the influence mechanism of nanoparticle volume concentration on flow performance of intercooler at constant heat transfer rate by changing the coolant mass flow rate is analyzed. The gas turbine operates in a 100% condition. The coolant inlet temperature is kept fixed at 293 K and nanoparticle volume concentration is increased from 0% to 5%.

Fig. 3 shows the comparison of the mass flow rate of $Al_2O_3$-water and Cu-water nanofluids as a function of nanoparticle volume concentration. It can be seen that the mass flow rate of the two nanofluids increases with the increase of nanoparticle volume concentration at constant heat transfer rate. This is due to nanofluids having lower specific heat compared to the base fluid, which is shown in Fig. 4. Besides, the mass flow rate of Cu-water nanofluids is much more affected by the nanoparticle volume concentration because Cu has a stronger effect on specific heat than $Al_2O_3$. For instance, the mass flow rate of Cu-water nanofluids can increase 28.93% compared to the base fluid when the nanoparticle volume concentration is 5%. But the mass flow rate of $Al_2O_3$-water only increases 6.19% under the same conditions.

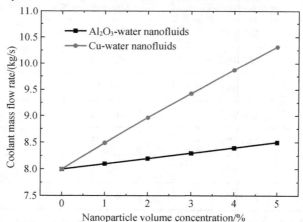

Fig. 3  Effect of nanoparticle volume concentration on coolant mass flow rate at constant heat transfer rate

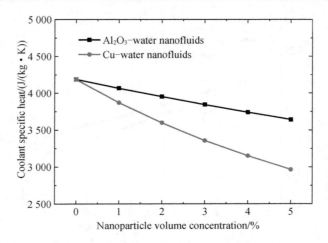

**Fig. 4  Effect of nanoparticle volume concentration on coolant specific heat**

Considering the effect of nanoparticle volume concentration on the density of nanofluids as shown in Fig. 5, the volumetric flow rate of nanofluids can be calculated using Eq. (30) and presented in Fig. 6. It is found that the volumetric flow rates of the two nanofluids are lower than that of base fluid, which is beneficial to reduce the pumping power. Besides, the volumetric flow rates of the two nanofluids is decreased with increasing nanoparticle volume concentration and Cu-water nanofluids have lower volumetric flow rates compared to $Al_2O_3$-water nanofluids due to higher density of Cu.

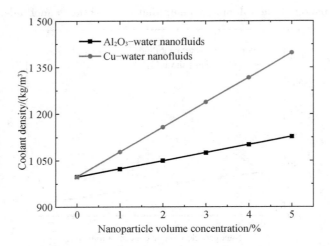

**Fig. 5  Effect of nanoparticle volume concentration on coolant density**

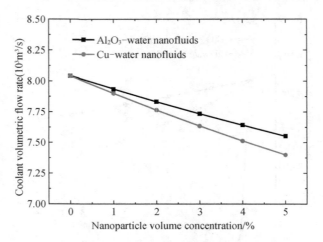

**Fig. 6 Effect of nanoparticle volume concentration on coolant volumetric flow rate at constant heat transfer rate**

When the mass flow rate and viscosity of coolant are considered together, Fig. 7 shows the changes of the Reynolds number of $Al_2O_3$-water and Cu-water nanofluids as a function of nanoparticle volume concentration. From Fig. 7, we can see that the Reynolds number of $Al_2O_3$-water nanofluids is lower than that of the base fluid and it is decreased with the increase of nanoparticle volume concentration. However, the Reynolds number of Cu-water nanofluids changes in the opposite trend. The reason is that the ratio of mass flow rate and viscosity increases with the increase of nanoparticle volume concentration for Cu-water nanofluids.

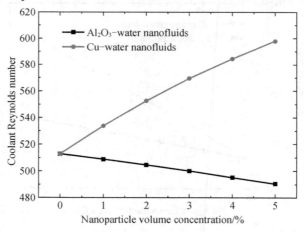

**Fig. 7 Effect of nanoparticle volume concentration on coolant Reynolds number at constant heat transfer rate**

Based on Eq. (26), the effect of nanoparticle volume concentration on the pressure drop of nanofluids is shown in Fig. 8. As shown in Fig. 8, it can be concluded that the pressure drop of two nanofluids is increased when the nanoparticle volume concentration increases at a constant heat transfer rate. Then, the pumping power of nanofluids can be calculated using Eq. (29) and presented in Fig. 9. The results show that the pumping power of the two nanofluids is lower than that of base fluid and they decrease with the increase of nanoparticle volume concentration. Also, the decrements of pumping power of two nanofluids are different. For example, the pumping powers of Cu-water and $Al_2O_3$-water nanofluids can decrease 6.42% and 2.08% respectively compared to the base fluid when the nanoparticle volume concentration is 5%. Therefore, it is concluded that more pumping power can be saved by using nanofluids and increasing nanoparticle volume concentration, especially for Cu-water nanofluids.

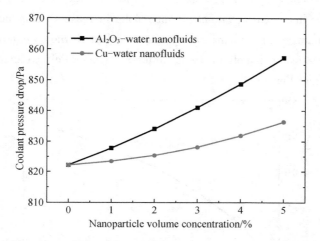

**Fig. 8** Effect of nanoparticle volume concentration on coolant pressure drop at constant heat transfer rate

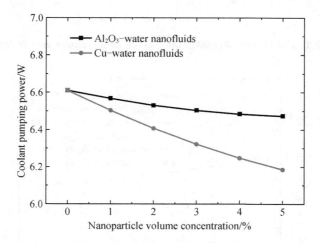

**Fig. 9** Effect of nanoparticle volume concentration on coolant pumping power at constant heat transfer rate

## Effects of coolant Reynolds number

Coolant Reynolds number plays an important role in determining the heat transfer performance of an intercooler. The gas into the high pressure compressor may be overcooled or overheated if the coolant Reynolds number cannot be controlled properly. In this section, the effect of coolant Reynolds number on flow performance of an intercooler at constant heat transfer rate by changing the coolant mass flow rate is discussed in detail. The gas turbine operates in a 100% condition. The coolant inlet temperature is fixed at 293 K and the mass flow rate of the base fluid is increased from 8 kg/s to 20 kg/s.

Fig. 10 shows the changing of heat transfer rate with the Reynolds number of the base fluid. It can be seen that the heat transfer rate of the intercooler increase exponentially with the increase of coolant Reynolds number. Then, the effects of the coolant Reynolds number on the mass flow rate and pumping power of nanofluids with different nanoparticle volume concentration are presented in Fig. 11 and Fig. 12.

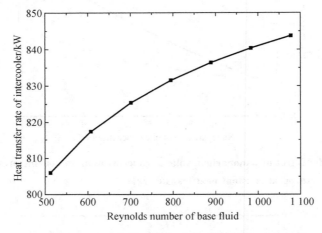

**Fig. 10  Effect of coolant Reynolds number on heat transfer rate of intercooler**

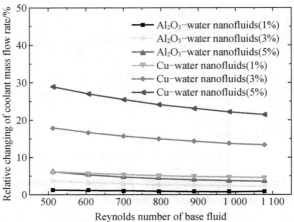

**Fig. 11  Relative changing of coolant mass flow rate with coolant Reynolds number at constant heat transfer rate**

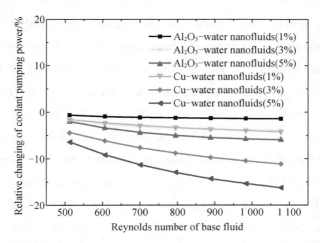

**Fig. 12 Relative changing of coolant pumping power with coolant Reynolds number at constant heat transfer rate**

As shown in Fig. 11, the effect of nanofluids mass flow rate will decrease with the increase of coolant Reynolds number, which means that the mass flow rate of nanofluids can be more affected by the coolant Reynolds number when the Reynolds number is relatively low. The main reason is that the thermal resistance ratio of gas side and coolant side can increase with the increase of coolant Reynolds number. Based on the results in Fig. 12, it can be concluded that the pumping power of the two nanofluids is lower than that of base fluid and the pumping power decrement of the two nanofluids increases when the coolant Reynolds number increases. For example, the pumping power of $Al_2O_3$-water and Cu-water nanofluids with the addition of 5% nanoparticle can reduce 2.08% and 6.42% respectively compared to the base fluid at 512 Reynolds number for base fluid. However, 5.93% and 16.31% reductions of pumping power for $Al_2O_3$-water and Cu-water nanofluids are achieved respectively at 1 077 Reynolds number for base fluid. This means that we can save more pumping power at higher Reynolds number by using nanofluids, especially for Cu-water nanofluids. Besides, both results show that the nanoparticle volume concentration is one of the important factors influencing the pumping power of nanofluids at constant heat transfer rate.

**Effects of coolant inlet temperature**

For the intercooler, the coolant inlet temperature can be influenced easily by the external environment, such as the temperature of sea water that will be changed with season and location. Meanwhile, temperature is another important factor influencing the thermophysical properties of coolant. Therefore, it is very necessary to study the effects of coolant inlet temperature on flow performance of the intercooler at constant heat transfer rate. The gas turbine operates in a 100%

condition. The mass flow rate of base fluid and nanoparticle volume concentration are fixed at 10 kg/s and 5% respectively. The coolant inlet temperature is increased from 278 K to 313 K.

Fig. 13 shows the changing of intercooler heat transfer rate with coolant inlet temperature by using base fluid. From Fig. 13, we can see that the heat transfer rate of intercooler linearing decrease with the increase of coolant inlet temperature. Based on the $\varepsilon$-$NTU$ method mentioned above, the effect of coolant inlet temperature on the mass flow rate of nanofluids is given in Fig. 14. A careful inspection of Fig. 14 reveals that the nanofluids mass flow rate shows a negligible chang over the inlet temperature range. For instance, the mass flow rate of $Al_2O_3$-water nanofluids can increase 4.59% when the coolant inlet temperature is 278 K. But the increment is only 6.27% when the coolant inlet temperature is 313 K. The reason is that the thermal conductivity of nanofluids as shown the Eq. (41) doesn't consider more effect of temperature except for the base fluid itself. Beside, it may be clearly observed in Fig. 15 that the coolant pumping power decreases with the increase of coolant inlet temperature due to the effect of temperature on viscosity. Fig. 15 also shows that the percentage of pumping power reduction is decreased with the increase of coolant inlet temperature for $Al_2O_3$-water nanofluids considering the effect of nanofluids thermophysical properties. However, the pumping power decrement of Cu-water nanofluids maintains nearly constant over the coolant inlet temperature range.

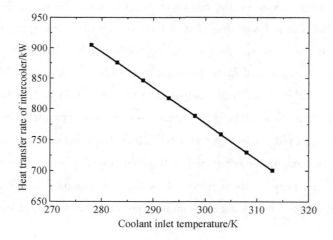

Fig. 13  Effect of coolant inlet temperature on heat transfer rate of intercooler

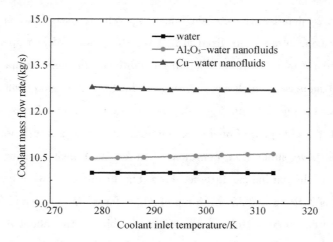

Fig. 14 Effect of coolant inlet temperature on coolant mass flow rate at constant heat transfer rate

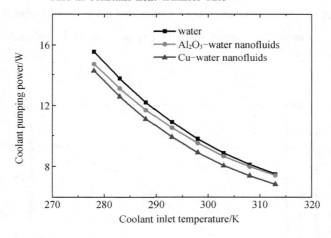

Fig. 15 Effect of coolant inlet temperature on coolant pumping power at constant heat transfer rate

According to the above analysis, it can be concluded that coolant inlet temperature is an important factor influencing the flow and heat transfer performance of intercooler. Compared with $Al_2O_3$-water nanofluids, coolant inlet temperature has no significant influence on the flow performance enhancement of Cu-water nanofluids by using the thermophysical property model as shown in this study.

## Effects of gas side operating parameters

In order to evaluate the flow and heat transfer performance of an intercooler by using nanofluids with different operating conditions of marine gas turbine, the flow performance of intercooler at different heat transfer rate by changing the gas side operating parameters as shown in

Table 2 is discussed in details. The coolant inlet temperature, mass flow rate of base fluid and nanoparticle volume concentration are fixed at 293 K, 10 kg/s and 5% respectively.

Due to the increase of gas turbine operating conditions, the inlet temperature difference of gas and coolant will increase, which can increase the heat transfer rate of intercooler as shown in Fig. 16. On this basis, Fig. 17 and Fig. 18 give the trends of nanofluids mass flow rate and pumping power with the changing of operating condition of marine gas turbine. The results reveal that both nanofluids have super flow performance at constant heat transfer rate with different operating conditions of the marine gas turbine. From Fig. 17 and Fig. 18, it can be seen that mass flow rate increment of 3.94% and 25.95% are obtained for $Al_2O_3$-water and Cu-water nanofluids under a 17% operating condition of the marine gas turbine. And about 4.73% and 9.01% decrease in pumping power for $Al_2O_3$-water and Cu-water nanofluids respectively. When the operating condition of the marine gas turbine is increased to 100%, 5.31% and 26.96% of mass flow rate improvement can be achieved for $Al_2O_3$-water and Cu-water nanofluids. Meanwhile, it is possible to save as much as 6.85% and 9.40% in pumping power by using $Al_2O_3$-water and Cu-water nanofluids.

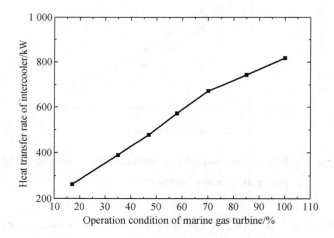

Fig. 16  Effect of operating condition of marine gas turbine on heat transfer rate of intercooler

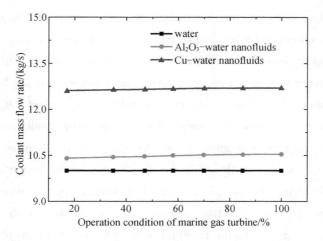

**Fig. 17 Effect of operating condition of marine gas turbine on coolant mass flow rate at constant heat transfer rate**

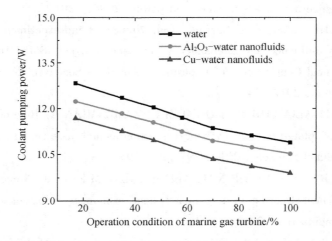

**Fig. 18 Effect of operating condition of marine gas turbine on coolant pumping power at constant heat transfer rate**

## Conclusions

This paper evaluated the application of nanofluids in intercooled cycle marine gas turbine intercooler based on effectiveness-number of transfer units ($\varepsilon$-$NTU$) method. Two nanofluids, $Al_2O_3$-water and Cu-water, were selected and used to analyze the effects of different parameters (such as nanoparticle volume concentration, coolants Reynolds number, coolants inlet temperature and gas side operating parameters) on the flow performance of intercooler at a constant heat transfer rate. On the basis of the work presented in this study, it is determined that the nanofluids have super flow performance by saving pumping power when the heat transfer rate is kept constant. More pumping power can be saved by increasing the nanoparticle volume concentration. When other conditions are constant, better flow performance can be noted for the nanoparticle with higher density and thermal conductivity and lower specific heat. Increasing coolant Reynolds number and decreasing the coolants inlet temperature are beneficial to improve the overall performance of intercooler by using nanofluids as coolant. It is also found that nanofluids can keep good performance under different operating conditions of the marine gas turbine and may be the new generation coolant for the intercooled cycle marine gas turbine intercooler. However, it is also worth noting that the thermophysical properties of nanofluids are important factors influencing the overall flow and heat transfer performance of the intercooler. Therefore, how to develop an alternative approach that is able to provide a quick solution to thermophysical properties prediction for nanofluids will be the key emphasis of study in the future.

**REFERENCES**

[1] WEN X Y, XIAO D M. Feasibility study of an intercooled-cycle marine gas turbine[J]. Journal of engineering for gas turbines and power, 2008, 130(2).

[2] WEI Dong, CHUN Mao, ZHU Jianjun, et al. Numerical and experimental analysis of inlet non-uniformity influence on intercooler performance [C]//ASME Turbo Expo 2012: Turbine Technical Conference and Exposition. American Society of Mechanical Engineers Digital Collection, 2012: 349-357.

[3] KHOSHVAGHT-ALIABADI M, HORMOZI F, ZAMZAMIAN A. Role of channel shape on performance of plate-fin heat exchangers: experimental assessment [J]. International Journal of Thermal Sciences, 2014, 79: 183 – 193.

[4] GUNNASEGARAN P, SHUAIB N H, ABDUL JALAL M F, et al. Numerical study of fluid dynamic and heat transfer in a compact heat exchanger using nanofluids [J]. Isrn Mechanical Engineering, 2012:1 – 11.

[5] CHOI S U S, EASTMAN J A. Enhancing thermal conductivity of fluids with nanoparticles [R]. Argonne National Lab., IL (United States), 1995.

[6] WEN D, LIN G, VAFAEI S, et al. Review of nanofluids for heat transfer applications [J]. Particuology, 2009, 7(2): 141 – 150.

[7] JANG, S P, CHOI, S U S. Effects of various parameters on nanofluid thermal conductivity [J]. Journal of Heat Transfer, 2007, 129(5):617-623.

[8] MURSHED S M S, LEONG K C, YANG C. Investigations of thermal conductivity and viscosity of nanofluids[J]. International Journal of Thermal Sciences, 2008, 47(5): 560 –568.

[9] MAHBUBUL I M, SAIDUR R, AMALINA M A. Latest developments on the viscosity of nanofluids[J]. International Journal of Heat and Mass Transfer, 2012, 55(4): 874-885.

[10] LI Y, TUNG S, SCHNEIDER E, et al. A review on development of nanofluid preparation and characterization[J]. Powder technology, 2009, 196(2): 89 – 101.

[11] SIDIK N A C, MOHAMMED H A, ALAWI O A, et al. A review on preparation methods and challenges of nanofluids [J]. International Communications in Heat and Mass Transfer, 2014, 54: 115-125.

[12] TRISAKSRI V, WONGWISES S. Critical review of heat transfer characteristics of nanofluids[J]. Renewable and sustainable energy reviews, 2007, 11(3): 512 –523.

[13] LEE S, Choi S U S, LI S, et al. Measuring thermal conductivity of fluids containing oxide nanoparticles[J]. 1999,121(2):280-289.

[14] XUAN Y, LI Q. Heat transfer enhancement of nanofluids[J]. International Journal of heat and fluid flow, 2000, 21(1): 58-64.

[15] XUAN Y, LI Q. Investigation on convective heat transfer and flow features of nanofluids [J]. Journal of Heat transfer, 2003, 125(1): 151-155.

[16] MURSHED S M S, DE CASTRO C A N. Superior thermal features of carbon nanotubes – based nanofluids-A review[J]. Renewable and Sustainable Energy Reviews, 2014, 37: 155-167.

[17] PANG C, LEE J W, KANG Y T. Review on combined heat and mass transfer characteristics in nanofluids[J]. International Journal of Thermal Sciences, 2015, 87: 49-67.

[18] HUMINIC G, HUMINIC A. Application of nanofluids in heat exchangers: a review[J]. Renewable and Sustainable Energy Reviews, 2012, 16(8): 5625 –5638.

[19] LEONG K Y, SAIDUR R, KAZI S N, et al. Performance investigation of an automotive car radiator operated with nanofluid – based coolants (nanofluid as a coolant in a radiator)[J]. Applied Thermal Engineering, 2010, 30(17-18): 2685 –2692.

[20] RAY D R, DAS D K. Superior Performance of Nanofluids in an Automotive Radiator[J]. Journal of Thermal Science & Engineering Applications, 2014, 6(4):041002.

[21] HERIS S Z, SHOKRGOZAR M, POORPHARHANG S, et al. Experimental study of heat transfer of a car radiator with CuO/ethylene glycol – water as a coolant[J]. Journal of dispersion science and technology, 2014, 35(5): 677-684.

[22] NARAKI M, PEYGHAMBARZADEH S M, HASHEMABADI S H, et al. Parametric study of overall heat transfer coefficient of CuO/water nanofluids in a car radiator[J]. International Journal of Thermal Sciences, 2013, 66: 82-90.

[23] HUSSEIN A M, BAKAR R A, KADIRGAMA K, et al. Heat transfer enhancement using nanofluids in an automotive cooling system[J]. International Communications in Heat and Mass Transfer, 2014, 53: 195-202.

[24] PEYGHAMBARZADEH S M, HASHEMABADI S H, JAMNANI M S, et al. Improving the cooling performance of automobile radiator with $Al_2O_3$/water nanofluid[J]. Applied thermal engineering, 2011, 31(10): 1833-1838.

[25] PEYGHAMBARZADEH S M, HASHEMABADI S H, NARAKI M, et al. Experimental study of overall heat transfer coefficient in the application of dilute nanofluids in the car radiator[J]. Applied Thermal Engineering, 2013, 52(1): 8-16.

[26] GNIELINSKI V. New equations for heat and mass transfer in turbulent pipe and channel flow[J]. International Chemical. Engineering, 1976, 16(2): 359-368.

[27] YOUSEFI M, ENAYATIFAR R, DARUS A N, et al. Optimization of plate – fin heat exchangers by an improved harmony search algorithm[J]. Applied Thermal Engineering, 2013, 50(1): 877-885.

[28] YANG S M, TAO W Q. Heat Transfer (Fourth Edition) [M]. Beijing: Higher Education Press, 2006.

[29] YUN H M. Study on flow and heat transfer characteristics of single phase in mini – channels[D]. Doctor thesis of Shandong University, Jinan, 2010.

[30] LEE S, CHOI S U S, LI S, et al. Measuring thermal conductivity of fluids containing oxide nanoparticles[J]. 1999, 121(2): 280-289.

[31] JUNG J Y, CHO C, LEE W H, et al. Thermal conductivity measurement and characterization of binary nanofluids [J]. International Journal of Heat and Mass Transfer, 2011, 54(9-10): 1728-1733.

[32] KOO J, KLEINSTREUER C. A new thermal conductivity model for nanofluids[J]. Journal of Nanoparticle Research, 2004, 6(6): 577–588.

[33] PRASHER R, BHATTACHARYA P, PHELAN P E. Brownian-motion-based convective-conductive model for the effective thermal conductivity of nanofluids[J]. 2006, 128(6): 588-595.

[34] KHANAFER K, VAFAI K. A critical synthesis of thermophysical characteristics of nanofluids [J]. International journal of heat and mass transfer, 2011, 54(19-20): 4410-4428.

[35] PAK B C, CHO Y I. Hydrodynamic and heat transfer study of dispersed fluids with submicron metallic oxide particles [J]. Experimental Heat Transfer an International Journal, 1998, 11(2): 151-170.

[36] XUAN Y, ROETZEL W. Conceptions for heat transfer correlation of nanofluids [J]. International Journal of Heat and Mass Transfer, 2000, 43(19): 3701-3707.

[37] HAMILTON R L, CROSSER O K. Thermal conductivity of heterogeneous two – component systems[J]. Industrial & Engineering chemistry fundamentals, 1962, 1(3): 187-191.

[38] BRINKMAN H C. The viscosity of concentrated suspensions and solutions[J]. The Journal of Chemical Physics, 1952, 20(4): 571-571.

[39] WEN C Z. Design and study on intercooling heat exchanger of marine gas turbine[D]. Shang hai: Shanghai Jiao Tong University, 2009.

# An Alignment Monitoring Device for the Output Shafting of Marine Gas Turbine[*]

WEN Xueyou, XIAO Dongming, DONG Bin, SONG Zigang, Kou Dan

(Harbin Marine Boiler and Turbine Research Institute)

## ABSTRACT

When a marine gas turbine is installed on the ship, the alignment among the power turbine, the output shaft and the reduction gearbox needs to be examined and adjusted precisely. After a long period of gas turbine operation, the alignment of the shafting needs be reviewed and adjusted regularly according to relevant regulations as well.

The paper introduces an alignment monitoring device, by which the alignment of the shafting under static and dynamic conditions can be acquired at any time. The device has been verified by real shipboard tests and the on-condition maintenance can be realized. The device can reduce significantly the workload of scheduled maintenance and can be used as a recorder for the responds of gas turbine against various dynamic shocks.

## Preface

The installation of the gas turbine on board is shown in Fig. 1. The gas turbine enclosure is installed on the hull base by use of shock absorbers, with an initial installation angle of α. The lengthways and crosswise rubber limiters are installed on both sides of gas turbine chassis.

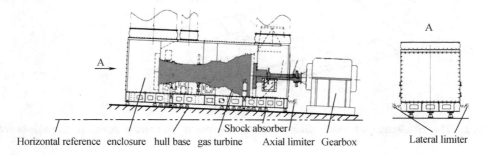

Fig. 1 The installation of the gas turbine on board

The marine gas turbine needs operating under the provided conditions, including longitudinal inclination, lateral inclination, longitudinal wiggler and lateral wiggler. Meanwhile, the gas turbine should be able to withstand the specified dynamic shocks.

---

[*] 本文为 2016 年美国机械工程师学会涡轮博览会会议论文。

During the long operating period, the alignment of the gas turbine output shaft can be changed due to the factors such as the deformation or the aging of the shock absorbers and the limiters. Re-examination should be carried out periodically. Usually, the connecting shaft flanges between the gas turbine and the reduction gearbox should be disconnected firstly. Then the alignment should be measured again and adjusted. While this reviewing method is time-consuming and may affects the normal operating of the gas turbine.

The alignment monitoring device for the output shafting of marine gas turbine presented in this paper can be used to observe and record the state of the gas turbine output shaft automatically. With simple structure, convenient operation and high accuracy, the device can reduce the workload and duty cycle of the scheduled gas turbine maintenance.

The alignment measuring of marine gas turbine output shaft is to mesure the posture of the output shaft in the 3D space. By comparing with the standard value, the deviation displacement and the parallel displacement can be obtained. At present, there is no similar product equipment.

The device can also be regarded as a test recorder used to record the response of the gas turbine against the dynamic shocks under the condition of unattended operation.

## The prototype and static test of the alignment monitoring device

The device prototype is simple extremely. The laser generators modified from the laser pens are installed on both sides of the engine chassis rear part. The laser generator on each side has vertical direction and horizontal direction, together with corresponding observation recorder(Fig. 2).

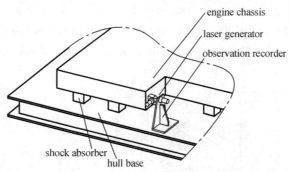

Fig. 2　The installation schematic diagram of the alignment monitoring device on the shipboard

The observation recorder can be operated visually on site or be installed with a continuous recording device. The conformity comparison between the measuring values by the observation recorder and the practical measuring values are carried out under various alignments of output shafting, on the platform of simulated marine gas turbine chassis in the laboratory.

## Real shipboard tests

Two sets of alignment device prototypes are installed on the starboard gas turbine chassis of the test ship(Fig. 3). The shock absorbers are installed under the chassis of marine gas turbine enclosure, while the connection between the reduction gearbox and the hull base is rigid.

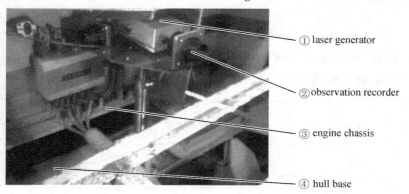

① laser generator
② observation recorder
③ engine chassis
④ hull base

**Fig. 3  The installation of the alignment on-condition measuring device on the shipboard**

## The relation between the deformations of the damper and the operating conditions under the straight line sailing state

Table 1  The relation between the deformations of the damper and the operating conditions

| Output Power | Idel speed | 10% | 20% | 35% | 40% | 60% | 80% | 90% |
|---|---|---|---|---|---|---|---|---|
| Deformations of shock absorber/mm | (Reference) | -0.11 | -0.16 | -0.21 | -0.24 | -0.28 | -0.37 | -0.43 |

## Steering test

In steering tests, the ship power remains unchanged. The change of the rudder angle causes the change of the course angle, and then affects the change of the power turbine(PT) output torque/speed and the navigational speed(Fig. 4).

In steering tests, the longitudinal inclination angle keeps unchanged, while the lateral wiggler angle has a fluctuation with larger amplitude. The changes of the longitudinal inclination angle and the lateral wiggler angle with the time are shown in Fig. 5.

The change of the lateral displacement of the shock absorber with the time in the steering test is shown in Fig. 6.

The deformation of the shock absorber is changed with the engine power and the output speed of the power turbine, as shown in formula(1):

$$\Delta Y = \frac{9\ 550 Ne}{n_{\text{PT}} W\ C_{tot}^{Y}} \tag{1}$$

In the formula  $\Delta Y$—The deformation of the shock absorber, mm;

$Ne$—The engine power, kW;

$n_{PT}$—The output speed of the power turbine, r/min;

$W$—The average distance between shock absorbers on two sides of chassis, m;

$C_{tot}^{Y}$—The sum of the elastic rigidity in vertical direction, for the shock absorber on left side (or right side) of the chassis, N/mm.

From the compartive analysis of Fig. 7, it can be seen that the theoretical calculation results are in good agreement with the practical measuring results.

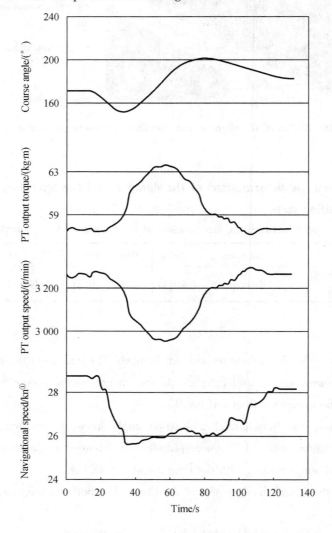

Fig. 4  Results from measurements on real ship

①1 kn = 1.852 km/h。

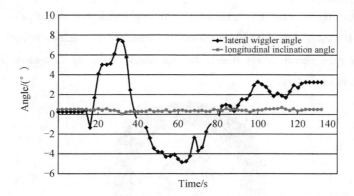

**Fig. 5** Changes of longitudinal inclination angle and lateral wiggler angle with time

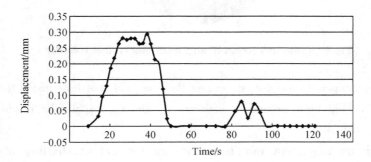

**Fig. 6** Change of lateral displacement of shock absorber with time

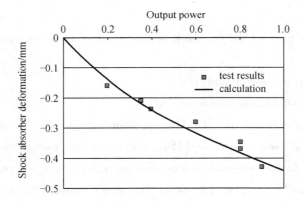

**Fig. 7** Comparison between theoretical calculation results and practical measuring results of shock absorber deformation

## The alignment state in the real shipboard test

There are three concentric circles and rings on the projection screen of the alignment observation recorder (Fig. 8). The projection reflects directly the alignment between the output of the power turbine rotor and input of the reduction gearbox. The laser point falling inside the circle

$A$ means that the shafting alignment is "excellent", falling inside the ring $B$ can be expressed as the "normal", which means that the alignment doesn't exceed the standard limit value while an attention should be paid deterioration trend; falling inside the ring $C$ means that the shafting alignment adjusting is necessary.

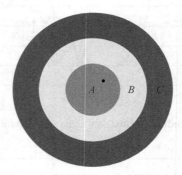

Fig. 8　Projection screen of alignment observation recorder

During the real shipboard test period, except that the light spot falls inside the ring $B$ for a short time when the ship turns around in high speed, the light spot falls inside the circle $A$ generally.

**Equipment design on alignment monitoring device for output shafting of marine gas turbine under dynamic and static state**

The schematic diagram of the alignment measuring device (AMD) is shown in Fig. 9, including:

(1) The measuring device (MD), which is composed of laser generators and the position solution unit;

(2) The information processing unit (IPU);

(3) The integrated information processor (IIP).

Through the concentrating lens, the laser is transformed into the light spot, which is projected onto the position sensitive device (PSD). Processed by the position solution unit, the information is uploaded to the integrated information processor. Now the alignment is shown on the screen of the integrated informaiton processor. The state for normal alignment is constant green light. The state for the trend of alignment being close to the limit is yellow flash at low frequency. The state for overrun alarm is red flash at high frequency and sound alarm.

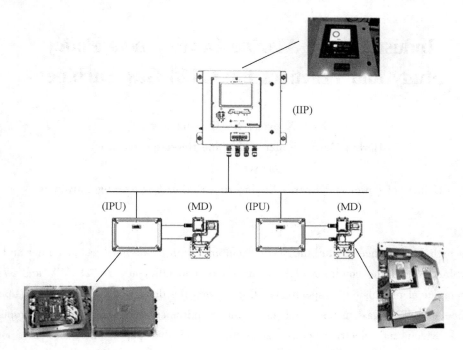

**Fig. 9 Schematic diagram of the alignment monitoring device(AMD)**

The device can show the alignment of the output shaft with static and dynamic display, and make a notice with colorful, sounded or light signals.

**Conclusion**

The device can display the alignment of the gas turbine output shaft in real time. With simple structure, convenient operation and high accuracy, the device can ruduce the workload and duty cycle of the scheduled gas turbine maintenance. The device can also be used as a recorder for the response of gas turbine against various dynamic shocks.

**REFERENCES**

[1]  WEN XueYou, XIAO Dongming, DONG Bin, et al. An alignment monitoring device for the output shafting of marine gas turbine[C]//ASME Turbo Expo 2016: Turbomachinery Technical Conference and Exposition. American Society of Mechanical Engineers Digital Collection, 2016.

[2]  XIAO Dongming, DONG Bin. A Real-shipboard test report on application of an alignment monitoring device (Principle Prototype) designed for the output shafting of marine gas turbine[R]. Research Report, HMBTRI.

[3]  SONG Zigang, LIU Xiaohua. An alignment monitoring device for the output shafting of marine gas turbine under dynamic and staic states [R]. Product Development Report, HMBTRI.

# Industrial and Marine Development Policy Study and Practices for GT28 Gas Turbine*

WEN Xueyou, XIAO Dongming

(Harbin Marine Boiler & Turbine Research Institute)

ZHAO Ningbo

(College of Power and Energy Engineering Harbin Engineering University)

## ABSTRACT

As a high performance gas turbine, GT28 combines with a two-spool gas generator and a free power turbine. Under the condition of ISO, its power and efficiency are 28 MW and 37% for marine mechanical propulsion, respectively. Considering the design characteristics and operating performance of GT28 gas turbine can meet the requirements of many marine propulsion, mechanical driven and electrical power generation, and this paper introduces the potential application of GT28 gas turbine in different industrial and marine fields. On this basis, the related key technologies are discussed briefly. Finally, a derivative network is presented to describe the relationships of different application and development of GT28 gas turbine.

## NOMENCLATURE

CC    combustion chamber
CCG    cross connect gear
CODOG    combined diesel or gas turbine
CODLAG    combined diesel electric and gas turbine
COGAG    combined gas turbine and gas turbine
COGAS    combined gas turbine and steam burbine
DLE    dry low emission
HPC    high pressure compressor
HPT    high pressure turbine
HRSG    heat recovery steam generator
IC    intercooled
ICR    intercooled and regeneration
IFEP    integrated full electric propulsion
LPC    low pressure compressor
LPT    low pressure turbine

---

\* 本文为2017年美国机械工程师学会涡轮机械技术会议与展览会论文。

LNG    liquid gas turbine
OTDF   outlet temperature distribution factor
PT     power turbine
RTDF   radial temperature distribution factor
S-S    water spray and steam injection
WTRC   wall temperature rise coefficient

**Preface**

In recent years, faced with the reduction of many energy resources and the deterioration of environment including global warming as well as pollution problems, hundreds of coal burning plants are phasing out by the energy industry. Energy conservation and emissions reduction have already become the common concerned problem in many countries. Besides, with the development of ocean resources and the attention of marine rights and interests, how to improve the performance of various ship equipment is a very important factor for defense security.

As a promising power machinery, gas turbine has been widely used in various industrial power engineering (such as mechanical driven and electrical power generation) and plays an increasingly significant role in the marine propulsion system due to its advantages of compact size, high power and less noxious emissions etc. However, considering the fact that gas turbine is a very complex equipment involving many advanced technologies, high innovative, heavy investment and long research period are the basic characteristics for developing a new type of high performance gas turbine. Therefore, gas turbine is usually considered as an important indicator to evaluate a country's industrial level, military and comprehensive national power.

Fig. 1 shows the market forecast information of industrial and marine gas turbine within the next few years[1]. It can be seen that GE Aviation plays a very important role in the fields of industrial and marine gas turbine. One of the major factor is the successful development of LM2500 series gas turbine, as shown in Fig. 2 (The datum in the figure are power and efficiency, respectively.). This may means that performance improvement based on the existing core engine is an effective way to derive different types of gas turbines.

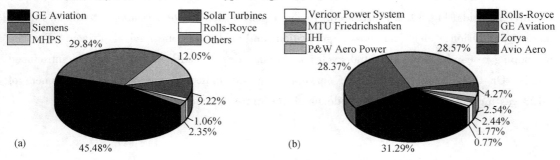

Fig. 1  Market forecast information of industrial gas turbine (2016—2025) (a) and marine gas turbine (2016—2030)[1] (b)

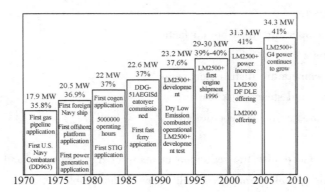

**Fig. 2  The development of LM2500 gas turbine**[2]

GT28 gas turbine, as illustrated in Fig. 3, combines with a two-spool gas generator and a free power turbine. Under the condition of ISO, the power and efficiency of GT28 gas turbine are 28 MW and 37%, respectively. Considering the performance of GT28 gas turbine can meet the requirements of many mechanical driven, electrical power generation and marine propulsion, how to realize the derivative development of GT28 gas turbine is a very important problem to extend its application range.

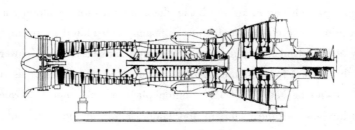

**Fig. 3  The basic structure diagram of GT28 gas turbine**

In this paper, many key technologies about the design and development of GT28 gas turbine in different fields(Fig. 4) are discussed. In the aspect of application research, it will be related marine propulsion, electrical power generation and industrial mechanical driven. For the developing research, three potential cycles including IC[3-4], ICR[5-7] and S-S are introduced briefly. On these basis, a derivative network about the industrial and marine development of GT28 gas turbine is put forward to understand the future directions of GT28 gas turbine.

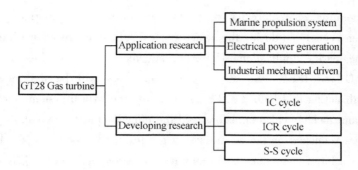

Fig. 4 Research scopes of GT28 gas turbine

## Marine and industrial derivative application strategy of GT28 gas turbine

According to the structural characteristic and operating performance of GT28 gas turbine, it can be used in the following fields.

(1) Warships. Destroyer, frigate, integrated electric propulsion module, etc.

(2) Civilian ships. LNG carrier.

(3) Electrical power generation. Self-provided power station, thermoelectric cogeneration plant, combined cycle power station, etc.

(4) Industrial application. Mechanical driven, natural gas transportation.

## The application of GT28 gas turbine on marine propulsion

### Overview

At present, gas turbine, diesel engine, steam turbine and nuclear power plant are the widely used main propulsion power plants of navy large and medium-szied surface ships. In this paper, three different combined power plants named CODOG, COGAG, COGAS and IFEP will be discussed mainly.

Table 1 lists the main performance parameters of several typical marine gas turbines for the mechanical drive under the condition of ISO. As shown in Table 1, GT28 gas turbine is somewhere in between LM2500 and LM2500 + from a performance perspective, which means that it can be applied in many destroyer, frigate and amphibious assault ship in the form of CODOG, COGAG, CODLAG and IFEP.

Table 1 Power and efficiency of different marine gas turbines

| Type | Power/MW | Efficiency/% |
|---|---|---|
| GT28 | 28.00 | 37.0 |
| LM2500 | 25.06 | 37.0 |
| LM2500 + | 30.20 | 39.0 |
| LM2500 + G4 | 35.32 | 39.3 |
| MT30 | 40.00 | 40.1 |

Since the operating characteristics of GT28 gas turbine as a marine power propulsion equipment has been considered during the design process, there are very few difficulties that need to be solved in the marine application. However, affected by the combined forms of CODOG, COGAG, CODLAG and IFEP, the transmission technology will be the important problem. Besides, it is worth noting that both the IFEP system and the hybrid system combing gas turbine power generation and mechanical propulsion need the effective power generation module.

In recent years, considering the advantages of gas turbine in compact size, high power, less noxious emissions, a lot of companies try to improve the performance of the LNG carrier by using the COGAS cycle.

### *Typical key technologies need to be broken through*

1) Cross transmission system

Considering CODOG has been widely used in various surface ships and the related transmission technologies were relatively mature, what should be paid more attention in the future is the cross transmission technology used in COGAG. Fig. 5 presents a schematic diagram of COGAG using 4 GT28 gas turbines. From Fig. 5, it can be found that with the application of the CCG transmission, any one GT28 gas turbine can drive the two propeller with equal power allocation at the same time. This leads that both operating power and efficiency of COGAG are improved obviously, especially for the low operation condition. Besides, there are many other characteristics for the COGAG shown in Fig. 5.

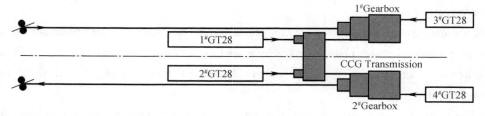

**Fig. 5   A schematic diagram of COGAG using 4 GT28 gas turbines**

### *For example*

(1) All the types of gas turbine are same, which means the related spare parts are easy to prepare. On this basis, the operation and maintenance of the gas turbine are relatively simple.

(2) No matter how many gas turbines are operating, every propeller can obtain the equal power from the gas turbine at the same time by using the CCG transmission system.

(3) To meet the requirement of different operating form of CCG transmission system, it is need to design many special clutch.

2) Power generation module based GT28 gas turbine

Recently, with the development of high performance warships, IFEP will become one of the most important developing trend. Unlike the traditional power system, marine integrated electric propulsion system includes generation, transmission, distribution, substation, drag, propulsion, energy storage, monitoring and power management. With the advantages of high power density, flexible modular arrangement, high integrated efficiency, fast start-up and the ability to be grid-tied, etc., gas turbine plays a very important role in the field of IFEP. Table 2 lists the related performance parameters of GT28 gas turbine under the condition of ISO when it is used for marine power generation module.

Table 2　Performance parameters of GT28 gas turbine used for marine power generation module

| Operating condition | | Environment temperature of 15 ℃, ignore the inlet and exhaust losses, light diesel oil |
|---|---|---|
| Outlet of electric generator | Power/MW | 25.6 |
| | Efficiency/% | 35.6 |

Due to the differences in load characteristics and system operating models, the following key technologies need to be solved when GT28 gas turbine is used for marine power generation module.

(1) Control technology. The equal power operating mode is usually seleted when GT28 gas turbine is used to drive the propellers. However, affected by the operating characteristic of generator set, the constant speed operating mode is applied when GT28 gas turbine is used for marine power generation module. Besides, considering the complex operating characteristics of warships(such as larger load range, quicker load change rate, etc.), marine IFEP makes strict demands on the control system in comparison to with the industrial gas turbine power generation system. Therefore, how to determine the operating state of the gas turbine and the fuel oil supply and control strategy are the important technologies to realize the secure and stable operation of gas turbine based IFEP.

(2) Mutational load adaptation technology. During the sudden load rejection process of gas turbine power generation module, the emergency vent valve installed in the casing of combustion chamber will be opened and deflated instantly. How to effectively avoid the overspeed and brennschluss and improve the adaptation ability of gas turbine are another important technologies.

3) Propulsion system of LNG carrier using GT28 gas turbine

Reliability, economy and low emission are very important indicators to evaluate the performance of the LNG carrier. At present, most of the LNG carrier select steam turbine as the main power engine because of its higher output power and better fuel applicability. However, the lower thermal efficiency and larger space requirement limit the development of steam turbine in the field of the LNG carrier. Attacted by the advantages of gas turbine in compact size, high power, less noxious emissions, both ocean operators of LNG carrier and engine manufacturers pay more attention on the gas turbine based propulsion system[8-9]. Fig.6 shows a schematic diagram of the combined gas turbine-steam generator electric and auxiliary diesel generator electric for LNG carrier. In recent years, many companies including Rolls-Royce, Alstom and GE have put forward the economical and efficient gas turbine based propulsion schemes for different LNG ships.

Fig. 6　Arrangement of gas turbine on the LNG carrier

Considering the initial design characteristics of the GT28 gas turbine, there are several key technologies that need to be solved when it is applied to LNG carrier.

(1) Combined gas turbine-steam generator electric technology. To reduce the operation cost and improve the performance of the propulsion system, the design and optimization of the combined gas turbine-steam generator electric according to the requirement of different LNG carrier is very important.

(2) Dual fuel technology. For the gas turbine used in the LNG carrier, it should take the boil off gas of the LNG and diesel oil as the fuel. This means that both the combustion system and the control system need to be adjusted according to the real requirement of the LNG carrier.

(3) LNG cold energy utilization technology. Different from the conventional fuel, LNG has high-grade cold energy. In the evaporation process of 1 t of LNG, about 250 kW · h cold energy can be obtained. Therefore, how to effectively use the cold energy is worth studying to improve the performance of the gas turbine.

(4) Fire safety technology. Considering the characteristics of inflammable explosion of LNG, it is necessary to enhance the fire safety of the gas turbine.

## The application of GT28 gas turbine to the industrial power plant

### *Overview*

According to the forecast data provided by Forecast International, it can be found that with world demand for electricity steadily increasing, gas turbines have gained prominence as clean-burning and cost-effective solutions. Besides, gas turbines, which for many years performed mainly peaking duties, are moving steadily into an ever greater share of the base-load and combined-cycle applications.

As shown in Table 3, the power and efficiency of COGAS power plant based on 1 GT28 gas turbine and 1 steam turbine are 33 MW and 46.2%, respectively. And the corresponding parameters can be easily added to 67.2 MW and 47% if another GT28 gas turbine is added, which can meet the requirement of most industrial power plant. Therefore, it is feasible to develop GT28 gas turbine in the fields of industrial COGAS power plant, as shown in Fig. 7.

**Table 3  Performance parameters of GT28 gas turbine used for industrial COGAS power plant**

| Operating condition | | Environment temperature of 15 ℃, ignore the inlet and exhaust losses, light diesel oil |
|---|---|---|
| Outlet of electric generator (1 gas turbine + 1 steam turbine) | Power/MW | 33.0 |
| | Efficiency/% | 46.2 |
| Outlet of electric generator (2 gas turbines + 1 steam turbine) | Power/MW | 67.2 |
| | Efficiency/% | 47.0 |

Fig. 7  Application of GT28 gas turbine on industrial COGAS power plant

### Typical key technology need to be broken through

1) Low emission technology

For the industrial power plant, one of the most important indicators is the pollutants emission. With the increasing strictness of international emissions standard, the traditional marine GT28 gas turbine used diesel oil cannot meet the emission requirement. Therefore, how to decrease the exhaust emissions is the first problem to realize the industrial application of marine GT28 gas turbine.

In recent years, Harbin Marine Boiler & Turbine Research Institute from China numerically and experimentally investigated the effects of gas nozzle, flame tube, import of pyramidal and air flow distribution of flame tube on the emission performance of the GT28 gas turbine. The detailed information can be found in Fig. 8 – Fig. 11, Table 4 – Table 6.

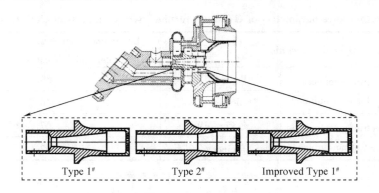

**Fig. 8  Different types of gas nozzle structure**

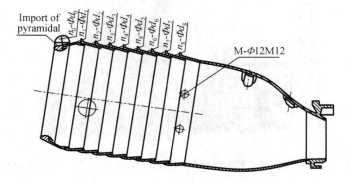

**Fig. 9  The structure diagram of flame tube**

**Table 4  The detailed information of flame tube**

| Type | Import of pyramidal Structure | Number of mixing holes | Film cooling holes | | | | | | | | |
|---|---|---|---|---|---|---|---|---|---|---|---|
| | | | $n_0 - \Phi d_0$ | $n_1 - \Phi d_1$ | $n_2 - \Phi d_2$ | $n_3 - \Phi d_3$ | $n_4 - \Phi d_4$ | $n_5 - \Phi d_5$ | $n_6 - \Phi d_6$ | $n_7 - \Phi d_7$ | $n_8 - \Phi d_8$ |
| Type 1# | | 6 | — | 65 - $\Phi$1.2 | 70 - $\Phi$1.2 | 70 - $\Phi$1.2 | 75 - $\Phi$1.2 | 75 - $\Phi$1.2 | 75 - $\Phi$1.2 | 75 - $\Phi$1.2 | 75 - $\Phi$1.2 |
| Type 2# | | 6 | 100 - $\Phi$1.5 | 120 - $\Phi$1.5 | 120 - $\Phi$1.5 | 150 - $\Phi$1.2 | 150 - $\Phi$1.2 | 150 - $\Phi$1.2 | 120 - $\Phi$1.2 | 120 - $\Phi$1.2 | 120 - $\Phi$1.2 |
| Improved Type 1# | | 4 | — | 65 - $\Phi$1.2 | 70 - $\Phi$1.2 | 70 - $\Phi$1.2 | 75 - $\Phi$1.2 | 75 - $\Phi$1.2 | 75 - $\Phi$1.2 | 75 - $\Phi$1.2 | 75 - $\Phi$1.2 |

**Fig. 10  Experimental platform of low emission combustion chamber**

Table 5  Experimental scheme of low emission combustion chamber

| No. | Gas nozzle | Flame tube |
| --- | --- | --- |
| 1[#] | Type 1[#] | Type 1[#] |
| 2[#] | Type 2[#] | Type 1[#] |
| 3[#] | Type 2[#] | Type 2[#] |
| 4[#] | Type 1[#] | Type 2[#] |
| 5[#] | Improved Type 1[#] | Type 1[#] |
| 6[#] | Improved Type 1[#] | Improved Type 1[#] |

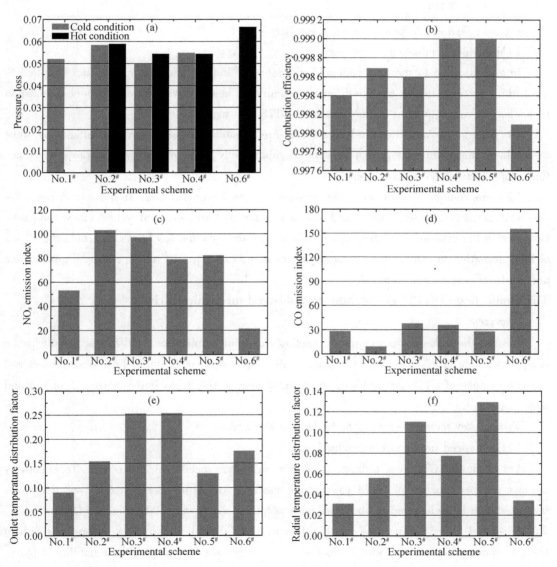

Fig. 11  Experimental result of combustion chamber with different factors

Table 6  Performance evaluation of different experimental scheme

| Parameters | No. 1# | No. 2# | No. 3# | No. 4# | No. 5# | No. 6# |
|---|---|---|---|---|---|---|
| Pressure loss | B | D | A | C | — | E |
| Combustion efficiency | B | D | C | E | E | A |
| $NO_x$ emissions | B | F | E | C | D | A |
| CO emissions | C | A | E | D | B | F |
| OTDF | A | C | E | F | B | D |
| RTDF | A | C | E | D | F | B |
| WTRC | F | E | C | B | D | A |

\* A and F represent the best and worst, respectively.

2) Multi-fuel combustion technology

In general, due to the environmental difference of industrial operator, the fuel supplied to gas turbine power plant are usually different. Therefore, how to develop the multi-fuel combustion technology is very important for the industrial GT28 gas turbine.

(1) Natural gas and diesel oil based dual-fuel combustion technology. To effectively solve the problem that nature gas cannot be supplied abundantly, it is necessary to improve the dual-fuel adaptability of GT28 gas turbine.

(2) New combustion technology for medium and low heat value fuel. In order to increase the economic benefit, various medium and low heat value fuel will be used in the many company. Considering the combustion characteristics of different medium and low heat value fuel, the effects of fuel properties on combustion chamber performance, flow capacity and operating life of turbine are worth investigating in detailed.

## The application of GT28 gas turbine on industrial mechanical driven

### *Overview*

Besides the industrial power plant, another important application field of gas turbine is the compressed natural gas transport. Considering the requirement of China's East-West project and the performance of GT28 gas turbine, the development of GT28 gas burbine become urgent and more important.

### *Typical key technology need to be broken through*

1) High-speed power turbine technology

For the initial GT28 gas turbine, its maximum output speed is 3 270 r/min. To effectively improve the performance of GT28 gas turbine used for industrial mechanical drive, the power turbine should be redesigned to obtain a higher output speed (about 5 000 r/min). On this basis, the high-speed power turbine shown in Fig. 12 can direct drive the natural gas compressor without the gear transmission system, which effectively reduce the cost and improve the reliability of plant. Besides, the increasing of power turbine speed is conducive to decrease the number of turbine stages and the outside diameters of the turbine blade. This means that it is necessary to improve the aerodynamic performance of the diffuser between low pressure turbine and power turbine.

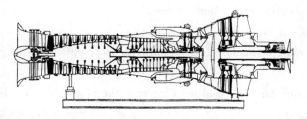

Fig. 12  GT28 gas turbine with high-speed power turbine

2) Low emission technology for nature gas based fuel

The related technology can be referenced the previously described content.

3) Integration and complete set design of gas turbine-natural gas compressor

Considering the complex operating characteristics of gas turbine and natural gas compressor, how to realize the performance matching of equipment, the integration of the system and the complete set design of the plant are very important, as shown in Fig. 13.

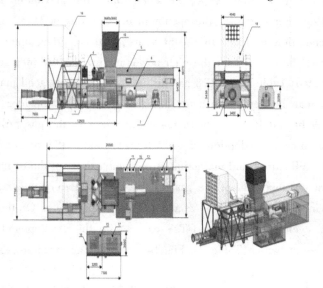

Fig. 13  A schematic diagram of gas turbine-nature gas compressor

## Marine and industrial derivative development strategy of GT28 gas turbine

In this paper, the following principles are considered to study the derivative development strategy of GT28 gas turbine.

(1) Significantly increasing the power.

(2) Increasing the efficiency.

(3) Making full use of the component aerodynamic performances of the initial GT28 gas

turbine.

(4) Making full use of the reliability of the initial GT28 gas turbine.

(5) Having the potential application demand.

(6) Easy carrying out the scheme.

After a large amount of analysis, the future development of GT28 gas turbine is mainly focused on three parts.

(1) IC cycle gas turbine. The goal is to increase the power 37 MW class.

(2) ICR cycle gas trubine. The goal is to increase the effciency of gas turbine, especially for in the part-load condition.

(3) S-S cycle gas turbine. The goal is to increase the power and efficiency using a simple method at a lower cost.

## IC cycle based on GT28 gas turbine

### *Overivew*

Analyzing the recent development of world marine gas turbines, it can be found that high-power (more than 22 MW) marine gas turbines have become a main trend of the marine propulsion power in the worldwide[10]. At the same time, the maximum power of marine gas turbines is still growing. For example, the maximum power rating of MT30 marine gas turbine is 40 MW under the condition of ISO. In recent years, with the development of the People's Liberation Army Navy, the demand for the high-power marine gas turbines grows.

As previously stated (Table 1), GT28 gas turbine is a relatively good marine engine with the power of 28 MW and can be applied to many different navy ships. Besides, Wen and Xiao put forward a sample way named IC cycle to increase the power of the gas turbine in 2007. Their analysis showed that after the adoption of an IC cycle under the precondition of performing minimum structure modifications and maintaining the compactness of the engine as a whole, there was still a significant enhancement of the gas turbine overall performance with its power output and efficiency being increased by about 34% and 4.1%, respectively. On the above basis, it can be found that the IC cycle based on the GT28 can realize the development goal of high-power marine gas turbine, as shown in Table 7. Fig. 14 presents the structure diagram of GT28 gas turbine and its IC cycle.

Table 7  Parameters comparison between GT28 gas turbine and its IC cycle

| Type | Power/MW | Efficiency/% | Turbine initial temperature/K | Pressure ratio |
| --- | --- | --- | --- | --- |
| GT28 | 28 | 37 | 1 270 | 22 |
| GT28-IC | 38 | 38 | 1 290 | 23 |

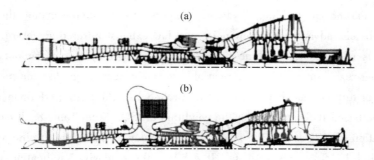

Fig. 14　Structure diagram of GT28 gas turbine(a) and its IC cycle(b)

*Typical key technology need to be broken through*

1) Compact intercooler and intercooled system

As an important part of the IC cycle marine gas turbine, the intercooled system including the intercooler and the sea water heat exchanger can directly influence the structure and performance of the engine. Considering the requirement of lower pressure loss and larger temperature drop in the air side of the intercooler, the compact design of intercooled system and the selection of coolant are very important. Besides, the safety of the gas turbine should be ensured even if the intercooled system is fault.

2) The performance matching of gas turbine components

The addition of an intercooler can change the thermal parameter of the high pressure compressor, combustion chamber and turbine. This means that the related components should be adjusted rationally or redesigned to realize the performance matching. Besides, the strength and vibration problems of IC gas turbine need to be paid more attention because the shafting of gas turbine will be extended.

3) Control technology of IC gas turbine

The existence of an intercooler can aggravate the time-delay characteristics of fluid flow and heat transfer, which make it difficult to match the thermodynamic parameters and develop appropriate control strategies for the gas turbine system, and may influence the maneuverability performance of ship. Besides, considering the addition of a new control parameter named intercooled degree, the control system of GT28 gas turbine is worth studying in detail to ensure the safe and stable operation of the engine.

It is also worth noting that due to the fact that IC gas turbine has not been applied in present navy ships, there are still many other key technologies which need to be further studied.

## ICR cycle based on GT28 gas turbine

### Overview

Besides the IC cycle gas turbine, ICR cycle is another effective way to improve the performance of GT28 gas turbine. The successful application of WR-21 marine gas turbine further verified the feasibility and effectiveness of the ICR cycle technology to increase the power and efficiency of the marine gas turbine. Besides, compared to the simple cycle, the corresponding ICR cycle has obvious advantages in low noise and low exhaust infrared signature.

On the basis of GT28 gas turbine based IC cycle, ICR cycle can be obtained easily by designing a proper regenerator and using a variable-geometry turbine guide vane instead of the first stage of the power turbine, as shown in Fig. 15. Table 8 lists the basic performance parameters of GT28 gas turbine based ICR cycle under the condition of ISO. From Table 8, it can be found that with the application of ICR cycle, the power and efficiency of GT28 gas turbine are increased by about 14.3% and 13.5%, respectively. It exhibits the extensive application prospect in the fields of marine mechanical propulsion and power generation module.

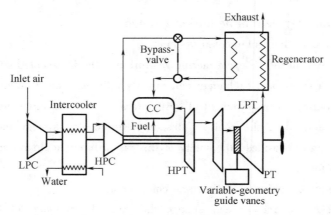

Fig. 15  Structure diagram of GT28 gas turbine based ICR cycle

Table 8  Parameters comparison between GT28 gas turbine and its ICR cycle

| Type | Power/MW | Efficiency/% |
| --- | --- | --- |
| GT28 | 28 | 37 |
| GT28 – ICR | 32 | 42 |

### *Typical key technology need to be broken through*

1) Effective and compact regenerator

As we all know, regenerator is the important components of ICR cycle gas turbine for heat exchanging between exhaust and fuel. Its performance will directly affect the efficiency of ICR cycle gas turbine. Considering the severe operating environment such as high temperature difference, high pressure difference and high-speed airflow disturbances, the reliability of the regenerator fin will be greatly challenged. Besides, the heat transfer performance of the regenerator is easy to decrease as the deposition of various exhaust pollutants, which may limits the development of ICR cycle gas turbine. Therefore, a reliable, high efficiency and compact regenerator is the first key technology that needs to be solved.

2) Variable-geometry turbine guide vane thechnology

The part-load performance is a necessary factor in the design of the marine gas turbine. To improve the operating performance of ICR cycle gas turbine, one effective way is to apply the variable-geometry turbine guide vane instead of the first stage of the power turbine. So how to improve the flexibility and reduce the leakage under the high temperature environment are worth paying more attention to.

## S-S cycle based on GT28 gas turbine

### Overview

To improve the performance of gas turbine, compressor interstage water spray and steam injection have been widely used in the fields of industrial power generation. In view of the potential application of GT28 gas turbine in industry, a new S-S cycle (Fig. 16) integrating the compressor interstage water spray and steam injection can be developed in the future, as shown in Table 9.

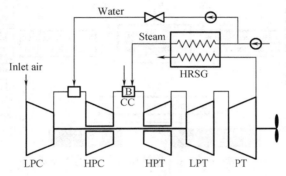

Fig. 16  Structure diagram of GT28 gas turbine based S-S cycle

Table 9  Parameters comparison between GT28 gas turbine and its ICR cycle

| Type | Power/MW | Efficiency/% |
| --- | --- | --- |
| GT28 | 28 | 37 |
| GT28-S-S | 31 | 39 |

### Typical key technology need to be broken through

1) Water spray between high and low pressure compressor

Considering the structural design characteristics of GT28 gas turbine, it is easy to implement the water spray between the high and low pressure compressor. In practices, the arrangement of the water nozzle, the spray performance, the related control strategy and performance matching of the gas turbine components caused by water spray are the basic key technologies to develop the S-S cycle gas turbine.

2) The overall design technology

To meet the requirements of the gas turbine in efficiency, reliability and security, the investigations into the flow distribution of water and steam, the integration and optimization of the water spray-steam injection system, the safety and life analysis of S-S cycle system need to be performed in detail.

Conclusions

This paper studied the industrial and marine derivative strategies of GT28 gas turbine from the perspective of a simple cycle based application and complex cycle based development, respectively. On this basis, the related key technologies are discussed briefly. According to the analysis in this paper, a derivative network (Fig. 17) is obtained finally to describe the relationships of different application and developments of GT28 gas turbine. It can be found from Fig. 17 that GT28 gas turbine has great potential in the industrial and marine fields (such as mechanical drives and electrical power generation).

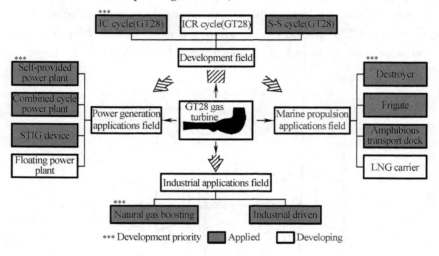

Fig. 17 Derivative network of GT28 gas turbine

Considering the design requirements and operating characteristics of various GT28 based gas turbines, the following common and key technologies need to be developed.

(1) Gas turbine power generation module for marine and industry.

(2) DLE combustion combustor and the combustion technology for medium and low heat value fuel.

(3) Cross transmission device and combining transmission device.

(4) Compact and efficient intercooler and the related intercooled system for high-power gas turbine.

(5) High speed power turbine.

(6) The overall performance matching technology for advanced cycle gas turbine.

## REFERENCES

[1] http://www.fi-powerweb.com/Industrial-Marine-Gas-Steam-Turbines.html

[2] http://www.ccj-online.com/oem-profile-aero-engine-portfolio-highlights-three-options-for-electric-generation/

[3] WEN X Y, XIAO D M. A new concept concerning the development of high-power marine gas turbines[J]. Ship Science and Technology, 2007, 29(4): 17-21.

[4] WEN X Y, XIAO D M. Feasibility study of an intercooled-cycle marine gas turbine[J].

Journal of engineering for gas turbines and power, 2008, 130(2):130 – 168.

[5] CRISALLI A J, PARKER M L. Overview of the WR – 21 intercooled recuperated gas turbine engine system: a modern engine for a modern fleet[C]//Turbo Expo: Power for Land, Sea, and Air. American Society of Mechanical Engineers, 1993, 78903: V03AT15A082.

[6] WEN X Y, LI W. WR – 21: a new generation of marine gas turbines[J]. Journal of Engineering for Thermal Energy and Power, 1999, 14: 1 – 6.

[7] XIAO D M, WEN X Y, ZENG X, et al. Exploratory study of a modification scheme incorporating intercooling and regeneration for a simple – cycle marine gas turbine[J]. Journal of Engineering for Thermal Energy and Power, 2004, 19(1): 89 – 92.

[8] ELLINGTON L, MCANDREWS G, HARSEMA – MENSONIDES A, et al. Gas turbine propulsion for LNG transports[C]//ASME Turbo Expo 2006: Power for Land, Sea, and Air. American Society of Mechanical Engineers Digital Collection, 2006: 65 – 76.

[9] BRICKNELL D J. Marine gas turbine propulsion system applications[C]//ASME Turbo Expo 2006: Power for Land, Sea, and Air. American Society of Mechanical Engineers Digital Collection, 2006: 85 – 94.

[10] FARMER R, DE BIASI B. Gas turbine world handbook[M]. Gas Turbine World, Fairfield, CT, 2010.

# 四、其他(船用燃机、船用柴油机)

# EGT 在燃气轮机燃用灰分燃料方面的经验 *

[法]M·莫瑞尔　J.P.贾仲耐　J.P.维维柯西

固定式燃气轮机在发电领域内肯定将起日益重要的作用。除经济方面的考虑外,尚有造成现今这一确定无疑趋势的一些技术原因:
1. 现代燃气轮机性能(功率和效率)的逐渐改进[1];
2. 燃气轮机应对日益严格的环境制约($NO_x$、CO、HC 排放)能力[2];
3. 燃气轮机在电力应用上的多方面适应性(调峰机组、基本负荷联合循环、热电联供装置);
4. 燃气轮机在燃料方面的灵活性等。

过去二十年间,全球石油资源呈现地区性集中,强烈地影响国际能源贸易。历次出现的"石油冲击"大大激发了消费者对低级油的兴趣。燃气轮机制造商抓住了这些动机的有利一面,发展了燃气轮机以灰分燃料运行的有效技术并验证这种发动机的可用性。

如今情况大有改观,天然气赢得了新电力项目中的极大部分的地盘,燃气轮机在燃用天然气的联合循环和热电联供装置方面已被推至前列。

但是在下世纪,能源的前景展望可能再次获得改观,这时煤气在 IGCC 装置中清洁燃烧将对电力供应提供一个长期的解决办法。在这种情况下,可用性和燃料灵活性都将保证燃气轮机在电力方面起着重要的作用。

实际上,尽管航空发动机及其派生发动机要依靠清洁燃气或馏出燃油运行,但重型燃气轮机可以使用种类广泛的燃料。
1. 轻质碳氢燃料(天然气、液化石油气);
2. 过程气体(焦炉气、炼油厂火炬气、煤伴生气);
3. 石油蒸馏燃料(煤油、柴油、真空蒸馏油、石蜡油);
4. 灰分燃料(原油、重油、混合渣油)。

---

\* 此文由闻雪友先生翻译,发表时间为 1994 年 9 月。

## 一、灰分燃料引起的问题

含金属的燃料对涡轮元部件可能产生的影响是多方面的。

1. 对燃烧系统的影响

(1) 燃料喷嘴被残碳物堵塞；

(2) 火焰筒因较高的燃气辐射率和在筒壁上可能结焦而过热。

2. 对涡轮及排气管的影响

(1) 因微粒撞击而磨蚀；

(2) 因灰分沉积而结垢；

(3) 因冷凝硫雾引起的冷端腐蚀等。

3. 对辅助设备(泵、流量分配器等)的磨损。

为了全面地掌握这些不利的因素，需要更深入地观察石油的一般化学性质和其在燃气轮机中的燃烧过程。更详细的叙述见附录Ⅰ和附录Ⅱ[①]。

## 二、石油产品的性质

原油是以碳、氢作为主要元素(两者比分别为 84%～86% 和 12%～14%)的有机化合物的混合物。由于其来源于矿物，显示出在分子规格及结构上的无限性以及微(痕)量元素($S, O, N, P, V, Ni, Co, Fe, Na, K, Ca, Mg, Al, Si, Cu$,等)的丰富多变性。

这些燃料 90% 以上是由以烷烃族表示的实际碳氢化合物所组成，范围从可溶的 $CH_4$ 到石蜡、环烷烃、芳烃和聚芳烃。

其他元素：氧、硫、氮、金属(钒、镍、铁等)虽然含量很低，但在原油和灰分燃油的一般化学性质中却是重要的。由于沸点高，在提炼过程中各元素大部分集中在"桶底"，使麻烦问题集中在重油上。

沥青烯具有高的分子质量(达 500 g/mol)，因此在电站进行输送时和在燃烧过程中均应严密注意。由于其在碳氢基油中苛刻的可溶性，并可絮凝和形成软的渣状悬浮物(图1)。此外沥青烯还显示慢燃的性质。

钒、镍和铁是沥青烯分子的核心元素。原油中钒浓度(质量浓度)变化从几个 ppm (1ppm 即 $10^{-6}$)(阿拉伯)到 $1\,200 \times 10^{-6}$(委内瑞拉)，在真空残渣中可超过 $5\,000 \times 10^{-6}$。除了某些原油(印度尼西亚、中国、委内瑞拉)外，镍的含量是较低的。铁不是原油的正常组分，可能被偶然地引入燃油。

次要的非碳氢形式是水和悬浮固体颗粒。乳化水通常带有以氯化物或硫酸盐形式的碱和碱土金属($Na, K, Ca, Mg$)。无机沉积物(沙、土、石膏、铁锈等)来自：

---

① 限于篇幅，附录Ⅰ和附录Ⅱ已省略。

1. 油田；
2. 燃油运送和储存装置；
3. 提炼过程（催化剂细粉）。

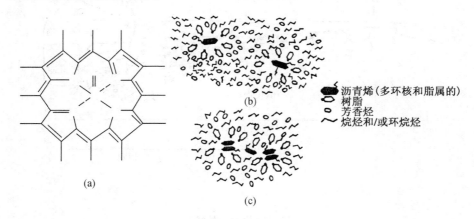

**图 1　燃油中的沥青烯**
(a)卟吩的结构；(b)沥青烯/树脂交互作用；(c)沥青烯分子间的相互结合

## 三、液体燃料的燃烧——复杂的科学

燃气轮机中产生的液体石油碳氢化合物的火焰完全不是简单的"$CO_2 + H_2O$ 的发生器"。首先，碳氢化合物燃烧不仅产生 $CO_2$ 和 $H_2O$，而且还有 $CO,SO_x,NO_x$ 以及未燃尽的碳氢化合物和烟炱。

碱金属（Na，K）生成相应的硫酸盐。这些硫酸盐在 2 200 ℃ 以下是稳定的，但熔点比较低，在过渡段内以液滴形式通过。

沥青烯和树脂的燃烧主要产生 $V_2O_5$，可能还产生 NiO。酸性的 $V_2O_5$ 的熔点很低，而且与硫酸钠一起生成易熔的盐。镍在火焰中氧化生成 NiO，但与 $V_2O_5$ 相反，NiO 表现为碱性并且不产生腐蚀性的副产物。

## 四、灰在热通道中的反应

热灰是非常活性的，可产生几种腐蚀威胁。热腐蚀主要是两种形式，均与熔盐的电化学腐蚀有关。

1. 硫化腐蚀。硫化腐蚀中的"入侵者"是碱金属和硫，它们形成低溶点的硫酸盐，硫酸盐溶解涂层，腐蚀超级合金。图 2 为严重硫化腐蚀过程的示意图。

2. 含钒易熔物的腐蚀。钒生成三种有害的低熔点化合物：$V_2O_5$（熔点 675 ℃）、钒酸钠和 $V_2O_5 - Na_2SO_4$ 低共熔物（图 3）。

腐蚀情况的详细讨论见附录Ⅲ①。

---

① 限于篇幅，附录Ⅲ已省略。

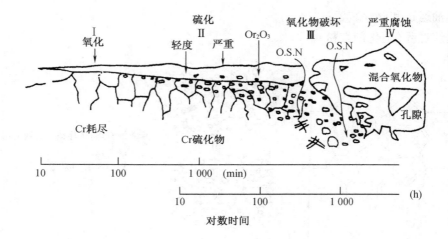

图 2　硫化腐蚀的机理

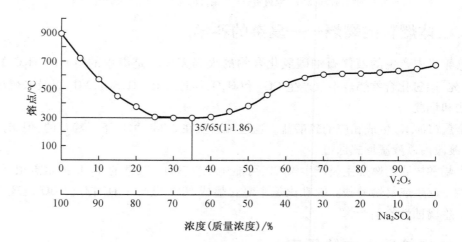

图 3　$Na_2SO_4/V_2O_5$ 混合物的静摩擦温度

## 五、EGT 控制腐蚀的基本观点

过去几十年间,由于美国 GE 公司的先锋作用被 Alsthom(现为 EGT,GEC ALSTHOM 和 GE 的合资企业)很快地追随,一项考虑所有制约因素的燃用重燃油的燃气轮机新技术出现了。同时,由石油工业实践中派生的适当的燃油处理技术对充分地控制热腐蚀的风险是有帮助的。

过去几年间已经发展了一个通用的方法,包括各种防范措施。

1. 基座上热腐蚀防范措施:抗侵蚀/腐蚀超级合金和涂层;经改进的喷嘴和燃烧室设

计;限制燃烧温度;涡轮流道形状最佳化(防止因沉积引起过分的压降)等。

2. 基座外热腐蚀防范措施:涡轮清洗橇;燃油和水处理装置;污水处理装置;渣输送装置等。

## 六、用先进的冶金技术控制腐蚀

无论是航空或固定式燃气机,在热力动力学方面为提高效率所需的代价是提高布拉东循环的热源温度。处于严苛的热力、机械和氧化条件下的涡轮部件是研制工作的核心。因此,在超级合金冶金以及有关的涂层和空冷技术方面的重要改进一直并仍然是涡轮部件进展的先决条件。

用灰分燃料的重型燃气轮机场合则更富有挑战性,因为这些机器必须显示优良的抗热腐蚀及燃烧性能。在1950年,重型燃气轮机的燃气初温约800 ℃,到20世纪80年代已提高到1 100 ℃,现在 EGT/GE 的 F 级燃气轮机初温为1 260 ℃。在涡轮动叶和导叶的材料方面一直进行着特殊的努力,既定的目标是在强度特性和抗热腐蚀性间做最佳的折中。

关于冶金技术最新发展水平的更详细情况见附录Ⅳ①。

## 七、严格控制燃料质量

硫蚀的危险通过下述措施控制。

1. 涡轮元部件的冷却及限制最高燃气初温。

2. 严格控制燃油中的碱金属浓度(Na 和 K 质量浓度小于 $1 \times 10^{-6}$)。如果必要,可用水洗燃料的方法来达到。有时也用铬基添加剂。

当钒含量(质量浓度)超 $0.5 \times 10^{-6}$,为防止钒腐蚀需用专门的抑制剂。在这方面,镁(Mg)化合物很早就一直被确认是一种有效的"灰分改性剂"。在火焰的下游,Mg 与 $V_2O_5$ 以一种类似于 Na 的方式发生反应,生成各种不同的"盐"或"混合氧化物",即

(1)$MgO + V_2O_5 \longrightarrow MgV_2O_6$(亚钒酸镁,熔点为740 ℃)

(2)$2MgO + V_2O_5 \longrightarrow Mg_2V_2O_7$(焦钒酸镁,熔点为930 ℃)

(3)$3MgO + V_2O_5 \longrightarrow Mg_3V_2O_8$(正钒酸镁,熔点为1 070 ℃)

$MgO/V_2O_5$ 平衡图如图4。

---

① 限于篇幅,附录Ⅳ已省略。

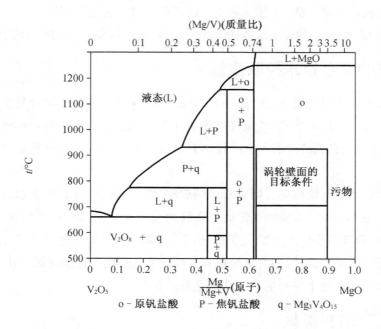

图 4 MgO/V$_2$O$_5$ 平衡图

当有足够过量的镁时,镁可以取代易熔的钒酸钠中的钠生成高熔点的、无害的盐(硫酸镁和正钒酸镁)。为了促进优先生成 Mg$_3$V$_2$O$_8$,镁的加入量必须超过相应于 Mg/V = 0.72(按质量)的反应理想配比值[3]。实际上,如果考虑到由于不可避免地生成硫酸镁而造成镁的消耗,必须注入相当于正钒酸盐理想配比所需镁量的 4 倍,最终的 Mg/V 大约是 3(按质量)。

除了少数的情况外,现在"3"这个比值已被广泛应用,因为其已被证明是充分控制钒腐蚀和限制灰分结垢的最佳剂量,因此,该比值已成为"人们认可的抑制值"。

## 八、一体化的燃油处理装置

燃油处理装置用于完成燃气轮机用燃油的基本处理。

1. 把燃油加热到适当温度,以降低燃油黏度;
2. 滤除粒渣;
3. 洗除碱土和碱土金属;
4. 加镁基添加剂以抑制钒蚀。

按照燃油处理流程(泵送、水洗和燃烧室中空气辅助雾化),在各段有不同的黏度限制规定。当储存有沥青烯絮凝风险的"敏感"燃油时,必须避免过度加热和搅拌掺气。

过滤是一项紧靠着燃油处理装置的关键处理,因为微粒会引起各种潜在问题:可能损坏输送泵和低压泵的叶轮,高压泵的螺杆以及流量分配器的齿轮;可能侵蚀燃油喷嘴,使燃油雾化场发生畸变;过量的渣末还可能过早地引起油水分离器的堵塞;可能由于某种稳定油/水乳化物的能力而影响油水分离。有必要强调滤器所起的"监督"作用,很遗憾,这个功能常常未被充分认识。

水洗时,碱和碱土金属通过均匀混合过程被转入水相,经稀化分散后的带盐水或借离心力去除,或借重力用静电力先凝聚成微水滴然后去除。通常需要多级分离。用低浓度(质量浓度)的破乳剂($5\times10^{-6}\sim50\times10^{-6}$)来加速分离。这种高技术的表面活化剂吸附在液滴的界面上并使界面层脆化,同时也消解了原油的某些天然混合物的乳化效应。

在轻油、低钠沾染的原油或馏出油的场合,用一个离心分离器简单地脱水(或净化),通常足以得到符合规范的燃油。离心和静电分离器足以得到符合规范的燃油。离心和静电分离器的优缺点比较见表1。

**表1　离心和静电分离器优缺点对比**

| 参　数 | 静电分离器 | 离心分离器 |
| --- | --- | --- |
| 机理 | 静电 | 动力 |
| 控制系统 | 半自动 | 全自动 |
| 达到启动稳定的时间 | 4 h(每级),12 h(总计) | 总计 30 min |
| 可靠性 | 好 | 好 |
| 维护成本 | 很低 | 低(高速设备) |
| 维修间隔期 | 一年 | 数月 |
| 固体杂质去除 | 尚好(需要附加过滤) | 极好 |
| 除渣 | 用废水排放,但容器内可能沉积,大致每两年需清理一次容器 | 用废水排放,靠自动门打开 |
| 最高运行温度 | 140~150 ℃(在压力下运行) | 98 ℃(在大气压下运行) |
| 除 Na 和 K 的性能 | 如果燃油是在系统设计规定范围内,则符合规范 | 如果燃油是在系统设计规定范围内,则符合规范 |
| 燃用高黏度燃油时的性能 | 好 | 好(增加装置数量) |
| 废水中的含油量 | 典型的是 $500\times10^{-6}$<br>最大 $50\,000\times10^{-6}$(启动时) | 典型的是 $50\times10^{-6}$<br>最大 $1\,000\times10^{-6}$(除门打开时) |
| 当燃料不符合水洗燃料规范时,修正所需响应时间 | 几小时 | 几分钟 |
| 运行中使用的灵活性 | 系统不能部分停运 | 在连续运行中,如果安装有备用机,离心分离器可以从使用状态中退出 |
| 所需空间 | 需要大的面积 | 有限的面积 |
| 安装方式 | 户外 | 户内或顶棚下 |

表1(续)

| 参 数 | | 静电分离器 | 离心分离器 |
|---|---|---|---|
| 噪声 | | 极低(静的) | 高(高速) |
| 水混合 | | 剪切阀(小,便宜) | 滞留低速混合器(大,贵) |
| 电耗 | | 低 | 正常 |
| 成本 | 小系统<20 t/h | 较高 | 较低 |
| | 大系统>20 t/h | 较低 | 较高 |
| 最佳处理方式 | | 连续 | 间断 |

图5、图6分别示出了基于离心和静电分离器的典型的燃油处理装置。

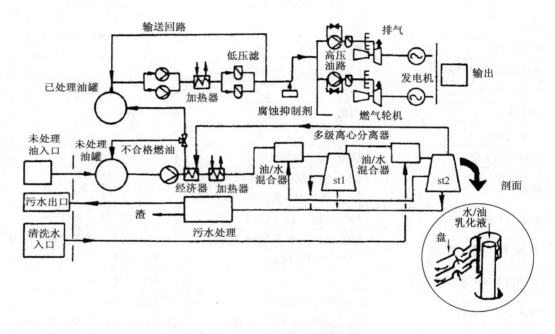

图5 离心分离燃油处理装置

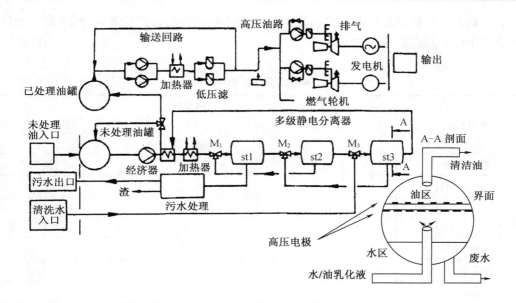

**图6 静电分离燃油处理装置**

现在,镁添加剂采用加药泵在线添加。目前,使用的是油溶性镁抑制剂而不是水溶性镁抑制剂(硫酸镁)或镁悬浮物(氢氧化镁),因为前者有明显的优点,如高纯度、完全混合的能力、没有磨损破坏等。

## 九、EGT 燃用灰分燃料经验之概述

1961 年 EGT 发运两台 Frame3 机组到阿尔及利亚,1972 年又有两台,均以未处理的原油运行,现在已累积运行了几十万小时。

此后,EGT 在世界各地安装了大量的用重质燃料的燃气轮机电站。

(1)中东的卡塔尔、伊拉克(原油);

(2)北非的摩洛哥(重油);

(3)亚洲的巴基斯坦(混合重油)、中国(原油)。

表 2 中选录了三个最有代表性的燃用重油或原油的电站,并给出了总的累计运行时数。

**表2 燃用重油或原油的代表性电站**

| 国家/公用事业单位 | 电站 | 燃机型号/数量 | 燃油 | 除盐设备 | 总运行时间(至 1992.12)/h |
|---|---|---|---|---|---|
| 摩洛哥/ONE | 阿戛特 坦吉斯 坦吐恩 | MS5001P/6 | 重油 | 静电分离器 | 约 180 000 |
| 中国/HIPDC | 汕头1 | MS6001B/2 | 中国原油 | 离心分离器 | 约 50 000 |
| 巴基斯坦/WAPDA | 柯特阿都 | MS9001E/4 | 混合油 | 静电分离器 | 约 50 000 |

表3给出了几种原油和重油的主要性能,有些中国原油的特点是其含镍量(质量浓度)要比含钒量(质量浓度)高得多。

表3 某些灰分燃料的平均性能

| 电站地点 | 柯特阿都 | 阿戛特<br>坦吉斯<br>坦叶恩 | 汕头1 |
|---|---|---|---|
| 燃油类别 | 混合油 | 重油 | 中国原油 |
| 密度(15 ℃) | 0.923 | 0.942 | 0.905 |
| 黏度(cst) (50 ℃) | 120 | 140 | 70.8 |
| 黏度(cst) (10 ℃) | 18 | 15 | 13.5 |
| 残碳量/% | 8.5 | 6.7 | — |
| 沥青烯/% | 3 | 2.5 | 2 |
| 闪点/℃ | 120 | 87 | 25 |
| Na、K(质量浓度)/$10^{-6}$ | 50~80 | 50 | 5~25 |
| (质量浓度)$V/10^{-6}$ | 3~50 | 65 | 1~5 |
| 低热值/(kJ/kg) | 40 850 | 40 750 | 41 000 |

下述的讨论反映了EGT及其客户所获得的经验。试图给出"重质燃料"电站如何管理的初步方法。

为使这些装置安全运行需采取一系列能提供良好运行可靠性的预防措施:

(1)精确的工艺过程控制;

(2)涡轮和燃油处理装置的细微的维修计划;

(3)对整个装置(卸油站、油罐区、污水处理装置等)的清洁服务工作。

由于重油的点火特性差,燃气轮机先用柴油启动,然后切换至重油。停机前则先切换至柴油,以免喷嘴、油滤等被黏性燃料堵塞。

燃气轮机的维修按EGT的标准设置检查计划表(考虑了与燃用灰分燃料有关的"维修因子"),检查内容包括:

(1)燃烧室检查;

(2)燃气热通道检查;

(3)大修。

在涡轮零件上的积灰使涡轮导向器通流截面积减小,涡轮气动效率下降,压气机压比升高,最终将导致压气机喘振,因此需要周期性的除灰。

现场经验表明有两种互补的除灰方法(图7)。

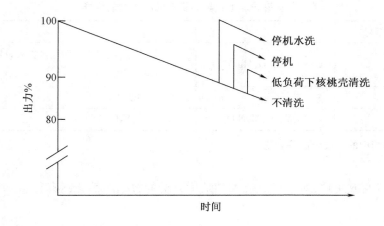

图 7　重油的清灰效率和清洗程序

(1)在发动机低负荷运行时喷入能轻度摩擦和易燃的材料(核桃壳清洗),可部分地恢复功率和效率。

(2)在发动机停机并冷却后,用高质量的热水来清洗,这几乎可保证发动机完全恢复性能。

应该指出,在停机、再启动的多次循环中涡轮有某种程度的自清洗能力。这是由于硫酸镁的天然吸湿特性。在停机期,沉积物中所含的上述盐吸湿,生成盐水化合物 $MgSO_4 \cdot xH_2O(x=6$ 或 $7)$,而发动机启动时盐水化合物经受突然脱水便成片剥落。

运行期间,在燃油管线的几个关键点上进行例行燃油分析。

(1)对未处理的、经清洗和加抑制剂的油进行例行的钠、钾分析,以保证正确地去除碱金属。在油清洗系统工作不正常时,不符合标准的油将转到未处理油罐中,直到系统恢复正常工作。

(2)分析钒、镍和镁主要是适用于加抑制剂后的油。亦建议对卸油处或未处理的油罐的油进行钒和镍分析,因为由此可以检查出供油方面的任何变化,可事先做出工艺调整。

(3)水和沉积物的含量对泵、滤器、污水装置等是重要的,在油处理装置的各部位进行分析。这种分析可提供在卸油站和储油罐的燃油的"清洁度"情况,和油水分离器的效率情况。

灰分的化学和结垢分析定期进行,以检验灰分的改性处理是否正确地进行,也提供了对积垢化学的更好了解。

表 4 示出了某些分析结果。观察到的化学图形显示了 $MgSO_4$,$MgO$,$Mg_3V_2O_8$ 等,这与前面的讨论相一致。从现场经验中也可了解或确认某些特点。例如,对灰垢的严格考察证实了一种分层的 $MgO/MgSO_4$ 结构。另一个例子是在含镍燃油中生成镍钒酸盐。

表4  从第一级上收集的积灰的结构图的标样

| X射线衍射信号 | A | B* | C |
|---|---|---|---|
| 主相 | $CaSO_4, 2H_2O, Mg_3V_2O_8$ | $MgSO_4, 7H_2O, NiSO_4, 7H_2O$ | $MgSO_4, 7H_2O, MgSO_4, 6H_2O, Mg_3V_2O_8$ |
| 中间相 | — | $MgNiO_2$ | — |
| 痕迹相 | $Ni_3V_2O_8, MgNiO_2$ | $Ni_3V_2O_8$ | — |

注:*指镍是燃油中的主要金属污染物。

无论是制造商或用户,对于建立燃油和结垢分析的数据库都会饶有兴趣。

## 十、结论

EGT在燃气轮机燃用灰分燃料方面的基本观点是基于数十年成功的经验。观点的核心是用前述的燃油清洗和抑钒方法严格控制燃气热通道中灰分的化学组成和化学性质。实际上,上述观点已为燃气轮机装置开辟了广泛的途径,据此,按燃用重质燃油进行设计的涡轮均可通过相应的燃油处理装置来输入燃油。

在质量不等的各种燃油条件下累积的80万运行小时表明:只要设计、运行和维修适当,这种动力装置是能够可靠运行的。

## 参 考 文 献

[1] BRANDT D E, COLAS M. A new advanced technotogy 50Hz gas turbine [R]. ASME International Gas Turbine & Aeroengine Congress & Exposition paper 90—GT 6338. 6/90.

[2] MOLLERE M. COLAS M. Gas turbines in urban areas:Water/Fucl Emutsions to reduce $NO_x$ emissions of gas turbines. GEC ALSTHOM Technical Review[J],1991,5:47-58.

[3] SCHILKE P W, FOSTER A D. PAPE J J. Advanced gas turbine materials and coatings [R]. NK:General Electric Company,1991.

# 苏联船用燃气轮机的摇篮
## ——МАШПРОЕКТ*

闻雪友

(哈尔滨船舶锅炉涡轮机研究所)

苏联作为一个海军大国,是世界上舰船燃气轮机的最大使用者。前不久,我有机会随代表团访问了位于乌克兰的舰船城——尼古拉耶夫市。这是一个宁静优美的城市,座落在黑海之滨,长期以来禁止外国人入内,现今也终于撩起了神秘的面纱,对外开放了。

苏联舰船燃气轮机的历史与机器设计科研生产联合体的发展息息相关,因为该联合体是在1954年为研制舰船燃气轮机之目的而建立的,是苏联唯一的一个船用燃气轮机的设计、研究单位。与其邻近的曙光生产联合体则是一个燃气轮机的批生产厂。

"'机器设计'加'曙光'等于从构思到生产的全部工作。""机器设计"的总经理罗曼诺夫说,"'机器设计'由设计、试验(部件与整机)、控制及试生产四部分组成,'曙光'厂在不同程度上参与了试生产,但归根结底,试生产是由'机器设计'联合体完成的。"

"两个企业原是苏联船用燃气轮机的垄断企业。历史已形成这样的局面:在苏联及现独联体的疆域内,水面舰艇的60%的动力是由我们生产出来的,并且还提供备件和监督服务。""曙光"厂的总工程师索洛金介绍说,"两个企业在学术上也很有名望,因为我们总是首先应用新技术,工程师、专家云集。此外,苏联科学院的许多专家为此服务,为企业研制了诸如电子束焊、激光、电化学、电物理等新设备、新工艺。"

"两个企业有不同的分工,但总的来说,很好地满足了海军的要求,也满足了民用工业方面的需要。"罗曼诺夫的话像是补充,又像是总结。

苏联最初的船用燃气轮机装置是 M-1 装置,功率 2.94 kW(4 000 hp),寿命 100 h,耗油率 557.8 g/(kW·h)(410 g/(hp·h)),M-1 是在航空发动机的基础上研制的。其后的 M-2 燃气轮机装置已是独立研制,寿命达 1 000 h,功率为 11 MW(15 000 hp),耗油率为 353.7 g/(kW·h)(260 g/(hp·h)),在各工况工作时在当时已具有最好的经济性,是苏联第一台全工况船用燃气轮机。

---

\* 文章发表时间:1993年5月。

在此期间,已确定了发展舰用燃气轮机的主要方向,许多方面保持至今:

(1) 直流布置。双转子套轴结构,可使尺寸小、质量轻、流阻损失小。

(2) 采用滚动轴承支承。

(3) 研制允许有较大不对中的联轴器。

(4) 发动机机匣与罩壳间用引射冷却空气来隔热。

(5) 采用双流路燃料空气喷嘴。

(6) 调节、操纵和保护系统。

(7) 垂直装配工艺和在专用导轨上发动机水平对接工艺。

(8) 海洋环境工作的进、排气装置,基座、罩壳、电子设备、润滑、启动、燃油和滑油系统。

在此期间所积累的经验甚至广泛地用于现代设计的舰船及工业用燃气轮机装置中。

1956 年,联合体分为两部分:

试验设计局——机器设计科研生产联合体;

批生产厂——曙光生产联合体。

1958 年为"乌克兰共青团员"号舰(北大西洋公约组织称之为"卡辛"级)研制了 M-3 主动力装置,功率为 26.5 MW(36 000 hp),由两台燃机并车组成,寿命 10 000 h,耗油率 353.7 g/(kW·h)(260 g/(hp·h))。该舰的特色是采用全燃推进,有可正、倒车的减速器,当时(1962 年)世界上还没有。英国类似的装置在 1969 年才用于 21 型和 42 型。美国驱逐舰"斯普鲁恩斯"号上的类似装置则是在 1973 年装舰使用。

1965—1966 年起开始研制第二代发动机,主要任务是提高经济性、寿命,改善声学特性。研究结果表明,舰船主动力装置应由不同功率的发动机通过连接系统组成一整体,可在任何航行工况下都能经济的工作。用这样的概念建成 M-5、M-6、M-7 装置,其中包括独立的、不同功率的巡航和加速机。

所采用的系统可保证在舰艇全速时全部装置投入工作,也可仅一台发动机工作,通过减速器将功率分配在两个螺旋桨上。在这些装置上应用了世界上第一次采用的倒车动力涡轮,快速作用的气动离合器等一系列新技术。M-5 装置用于"尼古拉耶夫"号舰("卡拉"级),M-7 装置用于"列宁格勒共青团员"号舰("克里瓦克"级)。

1971 年开始研究第三代燃气轮机:ГТД3000,ГТД8000,ГТД15000,效率 30%~35%。

ГТД3000 功率为 3 000 kW,三轴燃气轮机,简单循环,用作快艇和动力效应船的主动力,已累计运行 55 000 h。

ГТД8000 发动机功率为 6 000~8 000 kW,三轴,简单循环,用作快艇的加速机组及军、民用动力效应船,也是"光荣"号巡洋舰上的巡航发动机,船用型发动机已累计运行 $13 \times 10^4$ h。用于发电和天然气输送时功率为 6 000~6 300 kW。

ГТД15000 功率为 15 000~17 000 kW,三轴,简单循环,用作各种排水型水面舰艇的全工况或加速机组,也用作发电和天然气输送机组。与第二代燃气轮机相比,其燃气初温提高了 200~250 ℃,压比提高了近一倍。而且由于在高、低压涡轮上均采用了高应力单级涡轮代替两级涡轮,转子采用双支承,以及高性能冷却叶片、新材料、新工艺的应用等技术使发动机的比质量仍可接近航空发动机水平。基于这些发动机,对气垫船、水翼艇、水面舰艇建立了不同的装置,其中最有代表性的是排水量 13 000 t 的轻巡洋舰"光荣"号,其以一套燃-蒸联合循环装置(COGAS)作为巡航机,以提高巡航时的经济性,以大功率燃气轮机作为加速机。

从 1986 年起开始研究第 4 代船用燃气轮机 M80,目前正在研制进程中。

近四十年来,这个苏联唯一的船用燃气轮机装置研究单位发展了四代船用燃气轮机,共有 19 型燃气轮机装备了 26 种舰船,苏联 4 个舰队的所有舰用燃气轮机均来自该基地。

"机器设计"和"曙光"在舰船燃气轮机方面的顺利进展引起了热能动力专家们的注意。1966 年该单位获得了研制浮动电站的订货,1971 年第一艘自行式浮动电站"北极光"号(装有功率 12 000 kW 的 ГТТ-1A 机组)开往北冰洋沿岸。同时还建造了"灯塔"型列车电站,装有功率为 4 000 kW 的燃气轮机发电机组。1975 年开始研究 ГПА-10 机组,1979 年开始批量生产。第一套用于"雪比林克"压缩站,现今已用于 11 条天然气输送主干线上的 35 个压缩站。

在第三代发动机的基础上又发展了一系列用于动力工程和天然气增压的工业装置:

ГТТ-2.5(功率 2 500 kW,效率 28.5%)

ГТТ-6(功率 6 400 kW,效率 32%)

ГПА-6.3(功率 6 700 kW,效率 32%)

ГПА-16A(功率 17 000 kW,效率 35.5%)

ГТТ-16(功率 17 000 kW,效率 35.5%)

ГТТ-22(功率 23 000 kW(燃气轮机功率 17 000kW,蒸汽轮机功率 6 000kW);效率 43%)

ГТЭ-25С(功率 25 000kW,效率 45%)

最新的工作为回注蒸汽的 ГТЭ-25С 装置试验,ГТД25000(功率 25 000 kW)和 ГТТ-110(功率 $11 \times 10^4$ kW)燃气轮机发电机组的研制。

至今,由"机器设计"和"曙光"两个企业设计、制造的发动机和机组已超过 2 500 台,用于舰船、动力和天然气工业等。

机器设计科研生产联合体拥有一个强大的试验基地,有涡轮、压气机、燃烧室、轴承、传动、热电偶、应变测量、非标测量手段等六个部件实验室,约 20 个大型试验台架。

气动力学和热物理试验研究中包括大压气机试验台(驱动功率为 26.5 MW)、小压气机试验台(驱动功率为 2 MW),涡轮试验台,燃烧室试验台(6 台),叶片冷却试验台,风洞试验(进、排气装置,叶栅)。气源由四台燃气轮机驱动各自的压缩机通过串联或并联运行组成,最高压力可达 25 kg/cm$^2$,最大流量可达 70 kg/s。

当结束对苏联船用燃气轮机基地的访问时,我对其设计、试验研究能力,生产制造能力及规模,众多的船舶及工业燃气轮机型号,广泛的应用,新技术及高可靠性等方面留下了深刻的印象。